Out of the Fog

Tragedy on Nantucket

By Cindy Lou Young

Cover Art by Michael Mayne Russell, Kentucky
k.michael.mayne@gmail.com

Edited by   Nancy Grossman, Jeannie Tebbetts
and Wendy Becker

Published by Black Lab Publishing LLC
P.O. Box 64
Alton, New Hampshire 03809
www.bearandkatie.com

Printed by   Morgan Press, Inc.
60 Buckley Circle
Manchester, New Hampshire 03109-5233
www.morganpress.com

ISBN 09742815-9-X

Out of the Fog
First Print: March 2008

Published and printed in the United States of America

PRAISE FOR OUT OF THE FOG

"When I read Cindy's story, I was impressed by her courage to tell about the struggles she endured through her life. I was mostly touched by the story of her teenage years. The story is so informative that I began to think about it myself; what it must have been like growing up on a small Island thirty miles from the mainland. Teenagers everywhere were going to dances and concerts. It was during the years of a new age; 'Rock n' Roll' was the hottest thing. Not many people even heard of Nantucket; except through the book, 'Moby Dick.' It was an Island of locals whose families had settled there generations ago. Not the commercialized Nantucket of today. When Cindy became the miracle baby, everyone watched her grow up and everything she did must have been spoken about all over town. It must have been hard for her to live up to their expectations; the towns little miracle. And perhaps that is why she had so many obstacles to overcome. She may have been found in the fog that horrible day of the plane crash, but Cindy found her own way out of the fog. This is a great story and after I read it; I knew I wanted to be the one to publish it."

—Loni R. Burchett
Publisher/Black Lab Publishing, LLC

"I don't have to tell you – this is incredibly compelling reading. "

—Nancy Grossman
Editor, Backchannel Press

PRAISE FOR OUT OF THE FOG

"I received your manuscript today, and found myself entranced by the story..."

—**Don Costanzo**
Editor & Publisher
The Nantucket Independent

"I couldn't put the book down"

—**Vincent Greene**
New found friend

COVER ARTWORK

Created by

Michael Mayne
Russell, Kentucky
k.michael.mayne@gmail.com

ABOUT THE AUTHOR

Cindy Lou Young is a 51 year old wife, mother and Marketing professional. She was born on Nantucket Island on December 31, 1956. Cindy lived on Nantucket for 18 years before leaving to pursue her career. She now resides with her family in a small town in the Belknap county of New Hampshire, Gilmanton Iron Works. Gilmanton was the home of the well known Author of *"Peyton Place"*, Grace Metalious.

DEDICATION

FOR MY DAUGHTER

This book is dedicated to my daughter, Taryn. It was through the wonderful gift of being your mother that I came to realize the incredible bond that exists between a mother and child.

You have taught me many things about myself through the years. I am extremely proud to be your mom. Taryn, you are a free spirit and as you begin to spread your wings in life, I hope that you always carry that sweet smile and wonderful creativity in all that you do.

Remember to enjoy life; live it like you are dancing and nobody is watching! I love you.

ACKNOWLEDGMENTS

The author gratefully acknowledges the many contributors that helped with the creation of this book.

Loni Burchett for Publication.
Nancy Grossman for Editorial assistance.
Robert F. Mooney for historical accuracy.
Mark Laizure for digital imagery technical assistance.

The Nantucket Atheneum, the islands public library for research facilities and microfilm records of island newspapers.

The Nantucket Historical Association for its collections of materials related to the Northeast plane crash.

Bill Haddon for permission to reprint photographs

Paul Kozinn, Mrs. John J. B. Shea, James Driscoll, Frances Karttunen, Deborah Gore Dean and many others for sharing their personal experiences and memories of the crash.

JoAnn McHugh Richard for sending the actual parts of the plane that had been stored away in the attic by her mother. I am grateful for that "meant to be" moment!

Life Magazine, August 25, 1958, Tragedy for vacation-bound air travelers.

Newspaper articles, The Nantucket Town Crier, The Nantucket Inquirer & Mirror, The Standard Times, The New York Times and the Lewis L. Straus Papers

The National Transportation Safety Board, Civil Aeronautics Board Aircraft Accident Report, March 26, 1959.

The Nantucket Division of Probate and Family Court for Court documents.

All of the members of the local Message Board "YACK ON" for responding with memories about the crash.

SPECIAL THANKS

First and foremost I would like to extend my deepest appreciation to Loni Burchett and Black Lab Publishing, LLC for believing in this work and helping to make it a reality. Your tireless efforts and support have been completely overwhelming at times.

Thanks to my loved ones, Skip and Taryn. You have helped me throughout this project, moving me along with your love and support. Always know that you are my greatest gift in life and that I love you with my heart and soul.

Thanks to my girlfriends, Leona, Jen, Cheryl, Brenda, Diane and Wendy. You all have helped with your friendship and encouragement to keep me plugging. Even when I didn't think the end was possible, you showed me the path. I treasure our friendship and am forever grateful that you have been a part of my life!

Special thanks to Jeannie Tebbetts for editing support to get me started. Also, thanks to Nancy Grossman for the final edits. Our work together is just another example that God always sends me the help that I need.

My Day by Eleanor Roosevelt

AUGUST 22, 1958

I was much disturbed last weekend by the airplane crash on Nantucket Island. I know several people who spend the summer both on Martha's Vineyard and on Nantucket, and I was deeply grateful when one of my friends called to assure me he was not on the plane that had crashed. His plane had landed only a few minutes before the tragedy, and I'm sure he must feel he has been granted a new lease on life.

It is curious how difficult it is for most of us to realize that actually at any moment life may come to an end. It need not be on a plane. It can happen any place to any of us.

E.R.

FOREWARD

On August 15, 1958, a Northeast Convair bound for Nantucket with thirty-four passengers on board met disaster in a flash at the end of runway 24. Only ten survived and those who were there that night remember the selfless and heroic efforts of many.

Nantucket was a different place then. The community connection, the pace of life, and the landscape were vastly different from the island that exists today. Life was slower, the community smaller and the connections closer.

Nantucket was full of colorful New England "Yankee" characters, the "natives." Many have passed on or moved on and are no longer a part of the island community that exists today. They are only remembered by the remaining "islanders," characters like "Millie," "Brownie," "Frenchie" and others, unique personalities in a small community that definitely "watched out for each other."

Nantucket is now transformed into an affluent, resort-type island, with an atmosphere that is highly attractive to vacationers looking for a quaint escape by the sea. Development has had its impact on the fishing community. The landscape is now dotted with restaurants, t-shirt shops and boutiques.

The island community that I remember is long gone. However, there are still many old friends remaining on Nantucket who I long to see each time I return. It is for these memories and for those folks who were there for me and my family on Nantucket that I write this story.

It is time for this little piece of island history to be remembered and told by someone who lived it. Perhaps it is part of letting go. I think to let something go, you need to fully understand it. For so many years I have chosen not to dwell on this chapter of my life; on some occasions I have even chosen to ignore it. I can't explain why I now feel driven by some outside force to take a long, hard look at this, understand it and share it with others. I must admit that the more I learn and understand what took place, the more I find it amazing that I am even alive to tell about it. This is a tale of survival and heroism. I hope that readers will find it an interesting story too.

As I began this project, I was filled with ambitious thoughts of contacting as many of the survivors as possible. I wanted to learn more details about the crash, the victims, the survivors and the rescuers, through their own recollections of the event. My desire was to understand how this disaster impacted their lives.

This task, I soon learned, was close to impossible. Many of the people in this story have passed on or moved on to places unknown. I was forgetting that the other passengers were twenty or more years older than I.

One of the more interesting facts I learned from the few survivors and family members whom I was able to contact was that we all share one common, haunting, question: *"Why me, God?"* Each of us reflects on the horrific fate that many on Flight 258 suffered. We are left pondering, why was I spared and others were not? This is probably a common response with which most survivors struggle, but when I discovered this commonality between us, it really struck me.

In published accounts of the accident, I was referred to as the "miracle baby." I was the youngest of all of the survivors. Of the survivors, ten in all, I was the only person left unscathed, physically – I suffered only a scratch on my chin. It was the emotional scars that I carried with me throughout the following years.

Young children seem to attract attention when they miraculously survive traumatic ordeals. It was no different back in 1958. While I may not remember the actual details of the tragedy, or some of the people involved, I am the only person who can describe the impact and influence that this event had with regard to my own character. This book tells my story.

--Cindy Lou Young
2008

CHAPTERS

But for the grace of God....

Beginnings

It all started by accident. I was conceived by mistake; I certainly was not part of the "plan" for my parents. My father, Phil, was a sailor stationed at the Naval Base on Nantucket, a small island located thirty miles off the coast of Cape Cod, Massachusetts. A handsome twenty-year-old, he was outgoing, exciting and adventurous. Most probably it was his mystique that captivated my mother, a sixteen-year-old island girl who had grown up in this small, close-knit fishing community.

My mom, Jacqueline Anne Duce, "Jackie" as she was known, was a native of Nantucket. Her parents, my grandparents, Arnold ("Arnie") and Annie Duce had both come to the island with their families, Arnie's family emigrating from Latvia, Annie's from Holland. My dad's family lived in Troy, Ohio; he grew up on a 300 acre farm just outside of Dayton. His family worked the farm planting crops and raising livestock. Part of the attraction between my parents had to have been the sharing of stories about their lives that had been so completely different.

One night while babysitting, a night of love and passion turned quickly into a situation that neither of them had intended or expected. The end result: an unplanned pregnancy.

In the '50s, a teenage pregnancy meant shame and community rejection. Families usually dealt with "the problem" in one of two ways. The young mother-to-be was either sent to relatives who lived elsewhere for the duration of the pregnancy and put the child up for adoption, or a "shotgun" wedding would be quickly arranged for the bewildered young couple. No flash, no pomp and circumstance, just a usually small and simple ceremony and a pair of I do's

My dad's family was against the idea of marriage. They knew all too well that he wasn't ready for this type of responsibility. On the flip side, Annie and Arnie were urging my parents to wed, in the hope that marriage would salvage my mother's reputation in the community. Defiant, my dad went against his family's wishes, marrying my mom on Friday, September 15, 1956. The wedding was a simple affair, no extravagant arrangements. It was attended only by close family members and a few friends.

*

Three short months later, on New Year's Eve, December 31st, 1956, I was born, with no way of knowing that my mother and I would have only eighteen short months to share.

It wasn't until I had my own child that I realized how much bonding really takes place during those early days in our lives. It helped me to realize why I have this longing to know my mother. We had a connection. I just cannot remember it.

Before I was born, my dad transferred to a new assignment in Providence, RI. He was young, head strong and really rebellious. Having grown up on the farm, he was out to enjoy all of the excitement, temptations and adventure that life had in store for him. His free time off base was usually consumed with drinking and gambling.

The first year of the marriage was strained by lengthy separations. What had begun as a weak relationship grew weaker. My mom stayed on Nantucket, living at home with her parents until I was born. After my birth, we traveled to Rhode Island to be with my father for short periods of time.

We also spent time with my dad's family on the farm in Ohio; he would sometimes join us when he was granted enough leave to permit it. My dad loves to tell stories about the days when my mom came to visit on the farm. Because she had never experienced life on a farm, she was extremely naive when it came to many things that were just ordinary common sense to my dad's family.

My father had two brothers: Dave, the oldest, and George, the youngest. One time, George told my mother it was time to go out and brush the horse's teeth, so my mom grabbed the tooth brush, happy to take part in this task. My dad and his family still chuckle about that to this day.

My mother stayed in close communication with her parents while she was away from home. Her relationship with them was co-dependent; Annie and Arnie were both alcoholics. Alcoholism plagued her family. I can relate to some of the stresses that my mom must have faced during this period.

Annie constantly badgered my mom, trying to convince her to return to Nantucket. My dad's family commented over the years how Annie never gave my mom the chance to get comfortable living in Ohio and how much stress that placed on my mother. My Dad's mother, Grandma Young, used to say that my mom never had the chance to get close to their family, because Annie wouldn't leave her alone long enough.

As I began researching this story, I found several letters that were filed with court documents on Nantucket. I struggled, debating whether or not to include them in the story. I came to the decision that they should be included. They help show a bit of the family dynamics and the impact of alcoholism on our family. They also spell out the course of events leading up to the accident.

For me, the letters I found highlighted how, in an instant, a very active, full life can come to a sudden, screeching halt. Reading these letters was like watching the candle of my mother's life be abruptly snuffed out.

In July of 1957, just one month before the crash, Annie's closing note to her daughter was, "Dad and I are just lost without you, write soon."

This was followed by a letter from Arnie, in which he talks about their financial struggles and Annie's struggle with alcoholism. In closing was a PS. that referred to her starting over, once again, with A.A. She would make this attempt many times over, never finding the "gift" of sobriety. I often wondered if the accident left her faith so shaken that it might have become the barrier to finding her way to recovery. It certainly played a huge part.

Dear Jackie,

I thought I would write a few lines, how is everything going. Heard you and Cindy had quite a trip back. How is the weather out there? We have had terrible weather here ever since you left. It is getting better the last two days. We have been rabbitting the last few days – plenty of rabbits. Tex is just been going hog wild. The other day a rabbit jumped right over Tex's back. Boy did he go after him when he saw it. Too bad Phil couldn't have been here, boy what rabbits. I shot so many my gun barrel got so hot I had to put it in ice water to cool it off. And also the pheasants on the side.

Not much doing down this dam place yet. Mummie is promised a job, providing she goes back to A.A. down to Al Silva's garage (bookkeeping) a good job. Mummie has been worried stiff from not hearing from you. Jackie my Dear please write to her. Because it will help her a lot. She hasn't had a drink in ten days and me either. So please write to her. You know if she gets down she will hit it up again. So Jackie, it would please me. We would like to both get working so we can get out of debt again. Please excuse my hen scratching. I don't blame Phil's family for feeling the way they do towards Nana Lamens. And also Phil. She should mind her own asshole business, but she just can't do it. We have been staying with her ever since pop passed away. Boy what a life. How does Phil like his new job. Tell him to save up some dough and come down here for a few Blues. My boots will be dry by then.

Jackie if you do write don't mention about this letter. I don't want mummie to hear. It was so cold the other day it froze up the toilet up that god dammed shack. We watch T.V. every nite not a hell of a lot else to do. So Jack say hello to Mr. & Mrs. Young. So please do me a favor and write mummie and don't say a thing about this letter. So I will close now. Remember me to Phil and give Cindy a big kiss.

     Love Dad XXX

     PS. She went to A.A. tonite

Reading this letter brought back memories of my own child-hood. I remembered Arnie's role as the peacemaker, always bridging the divide between Annie and myself after each "bender." This disease was a constant battle for both Annie and Arnie, I am sure throughout their entire lives, and definitely throughout my childhood.

A familiar statement leapt off the page as I read this letter, the one where Arnie transfers the power or responsibility for his and Annie's sobriety to my mom. I heard this from Arnie many times, the insinuation that if I could somehow be good enough, or say and do the right things, the problems with drinking would go away.

The one thing that I have come to understand is that nobody has any control or responsibility for the actions of another. It took me a long time to learn and accept this, the opposite having been well instilled in my belief system from an early age. This is a huge realization for anyone involved in the cycle of addiction, whether it is substance or physical abuse. We have no control over any-one but ourselves.

Alcoholics and addicts think they have the power or ability to control others, which we do not. Recovery is all about finding a power outside of you. For me that power was God.

Many believe that alcoholism is an issue of willpower; it is not. I thought like this for many years. Addiction is a disease, just like diabetes. It is a physical addiction, compounded with a mental and spiritual affliction that can only be treated through connect-ing with others afflicted with the same suffering. Just like other physical diseases, it must be treated regularly. The only differ-ence is that the cure cannot be found in a pill or a liquid. It is only

attainable through the miracle of relating with other alcoholics
and making the important lifestyle changes as outlined in the
twelve steps of recovery.

The letter written by my grandfather refers to "Nana." This was
Annie's mother, my great-grandmother, "Nana Lamens," as our
family referred to her.

Nana, Dirkje (pronounced Dirk-ya), came to America with
her husband, Liendert (pronounced Leen-dert) or "Pop," from
Amsterdam, Holland. Prior to moving to Nantucket, they had
lived on Long Island in New York, where Annie was born. Nana
was a homemaker and Pop made his living hand-sewing fishing
nets. The small fishing village of Nantucket was the perfect fit for
their family.

What was probably the most difficult burden of all for my
mother was one that I, too, would come to know intimately – the
burden of being raised in a home affected by alcoholism. Arnie
and Annie's alcoholism had existed before the plane crash and
escalated to a heightened state afterward.

As I learned more about the crash, the experiences and details from those involved in the rescue efforts, I gained a clearer
understanding of the devastating impact that this event must
have had on both of their lives. I never understood this as a child,
or even in my early adult years; I only wondered why my life was
so screwed up.

As I read the letters, I also understood why my mother wasn't
writing or communicating with Annie. She was, most likely, escaping
the pain and reality of her life on Nantucket. This was something I
often did myself during my childhood. If there is any similarity
between us, my mom was most likely very disappointed and

angry with her mother. I had the same experience as a child. These feelings, stemming from the repeating cycle of never-ending disappointments, are a common experience in families caught up in the cycle of addiction, the perpetual empty promises of "this time it will be different," or "I will never do that again," or "no more for me, I will die if I keep this up." It struck me as I read these letters that my life was an eerie, strange repetition of many of the same circumstances that my mother had faced during her own brief life.

A few letters were exchanged prior to that August when my mother died, continued attempts by Annie to convince my mother to return home. While my mother was in Ohio with the Young family, my Grandmother Alma, and my Aunt "Babs" taught her many of the skills that she lacked. She had limited cooking skills, and knew nothing about the care of an infant. Even though she often found herself caring for her parents – and herself – during the drinking spells, her skills were born purely of survival rather than through parental nurturing and guidance.

There were so many simple things that my mother never learned by example at her own mother's side. Simple basic day-to-day tasks, like cooking, cleaning and doing the laundry, were foreign to my mother, who had lived most of her life caught up in the undertow of a dysfunctional, unpredictable and chaotic home.

Annie would swing periodically from obsessive house-cleaning to complete paralysis in her ability to maintain the home. These "swings" were usually the warning signs that another binge of heavy drinking could soon be expected. She often allowed the laundry to pile up into mountainous heaps, clean, but never folded or put away. Dishes were hidden in the oven, for days, without washing. Vodka bottles were stashed in any inconspicuous but convenient location.

*

I once heard alcoholism described as a persons' inability to find the "off" switch when they drank. I learned that it is much more. Alcoholism is a compulsion that takes over a person's life. Good decisions are quickly over-ruled by the addiction. Decisions are all based around the alcohol and the outcome leads to a life of insane situations and unpredictable behavior.

When I was growing up, one of the most popular television families was that of the Cleaver's, on *Leave it to Beaver*. One thing I knew very clearly – my family was nothing like the Cleavers.

My father once told me how he felt sorry for my mother when he caught his first glimpse of the reoccurring hell that was her home life. One day, shortly after they had met, he was with her when she came home to a scene that was all too familiar. As they walked through the door, to her embarrassment and disappointment, she was greeted by Annie in a drunken stupor. This was prior to the pregnancy, shortly after my parents had first met.

That day my dad witnessed my mom's deep, hidden rage that grew from the frustrations of the situation she was in. My mother began a frantic, systematic search of the house in an attempt to discover Annie's "secret stash." Another repeating pattern with

Annie, at the beginning of each binge, she would gather bottles of vodka, her drink of choice, and hide them all over the house – under the cushions of chairs, in the tank of the toilet, the freezer, the bottom of a trash can, or any other creative hiding place one might imagine. The creativity of the alcoholic is often a double-edged sword.

Every bottle my mother found, she smashed in the kitchen sink, crying the entire time, mortified to find herself caught up in her mother's hell once again. The cycle was repeating, the disappointment and the broken promises.

My dad brought his own mixed bag of issues into their budding relationship. He loved to gamble – that was his addiction. He followed his bouts of obsessive gambling with his own excessive drinking, nights that resulted in him spending or losing the little savings that he and my mom had put together.

On those occasions, my father was unable to provide sufficient financial support to sustain our family. At one point, my mother was forced to pawn her belongings to obtain food and pay for the basic necessities, the rent in this particular instance. She almost had to pawn a diamond ring, given to her by Joe Lennon, a family friend, as a wedding gift. Joe and Evelyn Lennon, who had been close friends with Arnie and Annie over the years, often helped bail my grandparents out of problems. Sometimes it was finding a place to live, other times, helping with their financial troubles. My mother had been able to keep from pawning the ring by giving it to my Grandmother Young for "safe keeping."

The reappearance of this ring many years later in my life was the triggering event that began my own journey to fill in the gaps, the many blank pages in my life. It was the force that motivated me to reconnect with the missing half of my family, the family that I did not know.

*

Annie continuously loaned money to my mother in a never-ending effort to encourage her daughter to come home. These loans were given partly out of love and concern and partly out of loneliness, but mostly out of guilt, Annie's way of trying to make up for the hurts caused by her alcoholism. In the end, dysfunction and the other stresses of my parents' forced marriage prevailed. The relationship deteriorated to the point of permanent separation. Eventually, petitions were filed for divorce.

My father initiated the first action. He filed for divorce on May 9, 1958, three months before the crash. His petition stated that he believed my mother was guilty of gross neglect of duty. It also stated that the details to support these claims would be presented at a scheduled court hearing in September, 1958. That day in court never came.

Arnie had responded by filing a petition on my mom's behalf; she was still a minor, not yet eighteen years of age. The petition asked that the court grant my mother parental rights and custody for my care. On June 5th, 1958 the court granted Arnie's petition. My custody was awarded to my mom.

Soon after the petitions were filed, in June of 1958, my mother and father decided to make one final, last-ditch effort to reconcile in Ohio. Annie sent my mother more letters during the months of June and July.

Also in July, Annie was admitted to Beech Hill, a rehab facility in Dublin, New Hampshire. In July, Annie wrote this letter addressed to us all:

Dearest Jackie, Cindy Lou and Phil:

Well I finally reached N.H. yesterday morning at 10:10AM. It is really a beautiful place. It is a farm house that has been modernized. They have 3 beautiful horses here, a mule, a goat and 3 dogs. They have a T.V. room and we play shuffleboard outdoors. I have a room with another girl named Rita. It's nice here, but home is better.

How was your trip to Ohio? I hope you made it ok and that everything is working out ok. I guess I will be here until August 8th, at least. So if you write here you will have to send it airmail. If not, then write to Dad and he can save the letter for me. I sure do miss you and Cindy Lou. I hope you will write often and keep us posted on things. If you take any pictures of Cindy, please send me the negatives. When I get home I will get your things together and send them to you.

Nana sure has been stinking since you left. I talked to her twice on the phone before I left Nantucket and boy did she rave and cry and carry on. She did not even think to say goodbye to me. Dad as much as told her to mind her own business. He told me to absolutely stay away from there when I get home.

As soon as I get home we are going to move. We have 2 or 3 prospects, so I hope one will work out. The horses they have here are trotting horses whatever that might be. Well honey it is lunch time so I will close. Please write soon and often and be sure that everything is ok.

By the way did you leave your diamond ring with Nana because Joe Lennon is on our tail about it? I love you and Cindy very, very, much and miss you more than I can tell. Write soon and give our best to Phil and Mrs. Young.

All my love, Mom XXXXXXXXX

Give Cindy a big kiss for Dad and I love you XX Mom

Nana Lamens is brought up again in these letters. Although Annie paints her in a very negative fashion, I remember her as a wonderful and caring woman. She would become a key person in my life during my early childhood years on Nantucket after the plane crash, someone whom I always felt "safe" with. An immigrant from Holland, Nana spoke broken English and fluent Dutch. I remember that when she got excited or upset, she could really go off on a tear in her native language. If you were around during these occasions, you would be left scratching your head, wondering, "what the heck is she talking about?" Regardless, you got the message, loud and clear, that she was definitely upset.

Using her Dutch was also a great way for her to keep me from hearing "adult" conversations. Discussions went on while I was right in the room. Not understanding Dutch, I had no clue what was being said.

Nana was also a wonderful cook; I can remember the great cookies she made around Christmas, and her odd habit of cutting up an apple to nibble on throughout the day. I still remember watching the pieces of apple sitting in the dish, growing browner and browner as they lingered over the course of the day.

She also had some quirky cures for whatever ailed you. A teaspoon of Vicks vapor rub on a spoon dipped in sugar was one I doubt I will ever forget. When I think of how many times she invoked this cure for a cold, I am amazed that my arteries are not permanently clogged with the goop. After any birthday party, there was also the ritual teaspoon of cod liver oil mixed in with a small glass of orange juice; this was needed, of course, to "keep you regular." I loved the fun of birthdays but always dreaded the cod liver oil "finale."

Annie's trip to Beech Hill was just one of many such trips to various detox centers in both Massachusetts and New Hampshire. There must have been thirty or more. The cycle of in-and-out admissions was a staple of my childhood. They began with either Annie or Arnie "falling off the wagon," the end result being another admission to the hospital or a detox facility. Annie made numerous trips to Beech Hill. Some facilities eventually gave up on her, refusing her further admissions after they had made so many fruitless attempts and efforts with her.

During one of her repeat visits to Beech Hill, she even managed to find a way to get her hands on alcohol. She got drunk – extremely drunk – right there in detox!. Their response was to send her home. During a hospitalization on Nantucket, she went from one patient's room to the next, drinking the mouthwash she found in an attempt to curb the effects of withdrawal.

Both Annie and Arnie tried to battle alcoholism with Antibuse, a medication that was supposed to make a person deathly ill if they drank while taking it. Both Annie and Arnie somehow managed to accept this result, finding a way to deal with the outcome – and continue their drinking. A local psychiatrist decided that perhaps he could help Annie, with prescriptions. He prescribed a variety of drugs in an effort to help her with depression and anxiety. Valium was one of them. Valium ushered Annie into another realm of addiction, one maybe even more powerful than alcoholism.

Arnie not only battled alcoholism, but suffered from manic depression as well, exhibiting periods of extremely unbalanced and strange behavior. These occasions were usually followed with an admission to the mental health hospital in Taunton, Massachusetts.

I remember the terror that consumed Arnie as he anticipated the dreaded "shock" treatments that had become a regular part of his treatment plan at the hospital. I only wanted him to get better, in the worst way. Although there were periods where he was able to manage the disease with lithium, he was never able to find true balance.

Arnie went through periods of avoiding taking his medication. I have learned over the years that this is common. The medicine calms the swings from emotional highs to emotional lows. To the patient, these swings are normal; life without them could seem boring. Arnie did not like being on medication and would often find ways to avoid it.

*

In early August my mother responded to Annie. This was her last letter:

Dearest Mom,

Sorry I haven't written sooner, but I didn't know your address. How have you been doing? It took me 3 days to get to Ohio and I was pooped. I called Phil when I got here. In fact, I almost ended up in Cincinnati by mistake, and Phil met me at his old apartment and I stayed there. Then we decided to take this place which is real cute. It sure is good to be back to him again.

I haven't heard about the episode with Nana. I really think that woman is off her rocker. She told me that I was a no good bastard and that I was nuts to go back to Phil. But, she kicked me out and if it wasn't for her, maybe Phil and I wouldn't have been separated in the first place.

Our house is real cute. We have [an] efficiency and it includes a kitchenette, living-room, twin bedroom and a bathroom with a shower. Everything is electric and we only pay $20.00 a week for everything. Cindy is doing just fine. I got her a pixie haircut today and you should see her, she's a living doll.

Well Mom, this is it for now except to say please write soon and we love you always.

Love always,
Jackie & Cindy

I knew nothing about this letter until I started researching material for this book. I can't describe the comfort and joy that came into my heart as I read my mother's words. It confirmed for me the love and the bond that I have always felt we must have shared.

It is unclear if the letter written by my mother was received before the crash. This next letter sent by my grandmother was dated just two days prior to the disaster. It is quite possible that the two letters could have passed each other in the mail.

August 12, 1958

Dear Jackie, Phil & Cindy Lou,

Dad and I are wondering what is wrong as we have heard nothing from you since you left. The only information we get is from outsiders. Is there any reason for your not writing to us?

Dad and I are awfully worried not hearing from you. I came home from N.H. last Wednesday. I wrote to you while I was there and sent the letter c/o Mrs. Young. Did you get it? Anytime you want to call us you can call Barbara (904) and she will call us to the phone.

I am working part-time for Reverend Brad Johnson. He is very nice. I do secretarial work. Don't forget to keep up on Cindy Lou's polio shots. We sure do miss her.

Dad and Nana had an awful row while I was away. She called Dad everything and ordered him out of the house. She has spread all over town that you called her names (you know what). I have not seen her since I came home. I certainly feel wonderful since I came home. It was quite an experience being there. I will write you about it later on. Will you please measure Phil for his sweater? Measure him from the armhole down for the length. I think I have reached the armhole, but I want to be sure.

Jackie, we received some papers from Dayton, Ohio, Juvenile Court saying the divorce was being heard on September 24, 1958. Dad wants to know what the story is. It also concerns support of Cindy. Please write us explaining what is going on. This is quite important.

I hope you and Phil are making out ok. Dad says to tell Phil that the bluefish have just reached Nantucket and hopes that the three of you can come here for a few days so Phil & Dad can go fishing. I am going to look at a four room apartment tomorrow. I hope we get it. We are going to a band concert tonight at Straight Wharf.

Well I have to go meet Dad now. We both send all our love to you, Phil & Cindy. Kiss Cindy for us. Please Write Soon or telephone.
All our love,
XXXXXXXXX
Mom & Dad

PS. I am ironing all your clothes and Cindy's. Do you want me to send them to you? Love Mom

At some point on August 13th or 14th, my mother contacted Annie and Arnie and made arrangements for us to return Nantucket. Annie wired money for plane tickets and related costs to my mother via Western Union and we set out on what would become the fateful trip back to Nantucket. There was no time to make reservations. My mother indicated to my grandparents that her plan was to fly to LaGuardia Airport in New York and either catch a connecting 10 PM flight to Nantucket, or go into the city and catch a train headed to Boston, then make other transportation connections to the island from there.

On Friday evening, August 15th, 1958, my mother and I boarded Northeast Airlines Flight 258 bound for Nantucket. The original 10 PM flight that she had intended to catch had been cancelled. Passengers scheduled for that flight were boarded as an extra section on flight 258. There wasn't much of a delay; the plane departed LaGuardia at 10:30. The 68-minute flight would be on the ground in Nantucket well before midnight.

In July of 1956, the year I was born, Nantucket experienced a memorable maritime disaster, the sinking of the *Andrea Doria*, which had collided with the Swedish liner, the *Stockholm*, not far off the coast of the island. Little did anyone realize that another disaster, the crash of flight 258, was just around the corner. Only a short year and a month after the sinking of the *Andrea Doria*, many more lives would be changed forever, including those of my mother and myself.

*For everything, absolutely everything, above and below,
visible and invisible... everything got started in him and
finds its purpose in him. – Colossians 1:16*

The crash

It was a hot, muggy August evening. Most of the thirty-four passengers that boarded Flight 258 at LaGuardia Airport were headed to Nantucket for a vacation. Trading the heat of the city for the island's ocean breezes was probably a most welcome thought for many of the passengers as the plane made its way through the late evening sky.

The island of Nantucket lies twenty-five miles south of Cape Cod and fifteen miles east of the Island of Martha's Vineyard. It is a small island, only forty-nine square miles in size.

*

The flight crew, Captain **John T. Burnham**, First Officer and co-pilot, **David C. Carey,** and stewardess **Arlene Dabek** from Manchester, New Hampshire, were busy running through the routine preparations for takeoff. Their departure was already about two hours behind schedule due to air traffic and ramp delays that had been encountered by a number of earlier flights that evening. At 10:25 PM, the twin-engined Convair CV-240-2 lifted off, bound for Nantucket.

At Nantucket Memorial Airport, a few workers were on duty at the ticket counter. Tom Giffin, the Senior Agent for Northeast Airlines, was working along with a few other maintenance workers. Cab drivers were parked outside the terminal awaiting the plane's arrival.

The once-sleepy little island airport was in the beginning stages of growth, driven by an influx of vacation visitors; the arrival of larger planes was just starting to become a regular occurrence. Motivated by this increased demand, the airport was in the process of lengthening its main runway, runway 24. White barrels marked a construction area at the end of the runway.

During the flight, personnel on duty at the terminal were in radio communication with the flight crew; there were also radio transmissions between flight 258 and Otis Air Force Base RAPCON (Radar Approach Control), which was located on the mainland in Falmouth, Massachusetts.

*

My mother and I were considered "standby" passengers, having boarded as part of the extra flight section. Due to our standby status, our names did not appear on the flight roster. This would result in quite a bit of confusion during the rescue, and hampered Annie and Arnie's efforts to confirm our presence on the flight after the crash.

The plane was not flying at its full capacity of forty; it carried only thirty-one passengers and the three crew members that night.

Some of the other passengers who had boarded Flight 258 with us included **Mrs. Joyce Bell**, a resident of Brooklyn, New York, **Margurite Boos**, age twenty-two, a resident of Astoria, Queens, New York, and **Edward Fennell**, age fifty, a sales manager from Morris Plains, New Jersey.

Gordon Dean, a fifty-two-year-old husband and father of two, was well known and respected for his work as the former head of the Atomic Energy Commission. Appointed to that position in 1949 by President Harry S. Truman, he served in this capacity from 1950 until 1953. Mr. Dean was at this time a financier in the field of private nuclear development. His wife was Mary Benton Gore, second cousin of Albert Gore, Sr., senator from the state of Tennessee. Mrs. Dean, who awaited her husband's arrival at the Nantucket airport that night, was one of the first responders to the crash.

Lita Levine, a young art student from Brooklyn, New York, was headed to the island for a summer vacation. Her parents Beatrice and Nathan Levine, back home in Brooklyn, awaited a call from Lita to let them know that she had arrived safely at her destination.

Eric Bannister, Jr., age twenty, a pre-med student attending Columbia College, was also from Brooklyn, New York. He was on his way to Nantucket to visit with friends. **Paul Kozinn** and **Don A. Breswick**, friends from the New York area, were headed to Nantucket to enjoy a short summer vacation.

Barbara Feinerman, another resident of Brooklyn, New York, and **Lawrence Stanley "Al" Foster, Jr.** a resident of Belmont, Long Island, New York, and **Barbara Dasch**, age twenty, of Bayville, Long Island, daughter of Dr. Joel Dasch a surgeon in Manhattan were also aboard. **Jeanne Cavin**, age forty-five, of Malvern, Pennsylvania, was an advertising director and long time employee of the *Friends Journal* in Philadelphia, a Quaker publication. Miss Gavin was heading to the island for a vacation.

Barbara Gillen, age twenty, a secretary employed by one of the television networks, was headed to the island for a visit with her friends, **Paula Haken**, age twenty-two, a television production assistant, from Jamaica, Queens, New York, and **Helen B. Hereford**, an employee of the CBS, in New York City.

William W. Hinckley, age forty-eight, a psychologist engaged in group therapy work at the Bleuler Psychotherapy group in Jamaica, Queens, New York, was headed to the island for his summer vacation.

Edward Kilberg of New York City had boarded this flight, as had **Sidney Rubenfeld**, a real estate broker, from Queens, New York, and **John P. O'Dell**, from Elgin, Illinois. **John S. Pearson**, an executive from New York City, was traveling to Nantucket along with his wife **Marylin Pearson**.

Virginia R. Marsh, age thirty, an assistant to an account executive at Benton & Bowles, an advertising agency, in Queens, New

York was aboard. So was **John J.B. Shea**, age thirty-three. A graduate of the Harvard Law School, Mr. Shea was a practicing attorney with the law firm of Cleary, Gotlieb, Friendly & Hamilton, 52 Wall Street, New York City, an active member of the Anti-Tammany faction of the Lexington Democratic Club at the time of the crash, he had served as a former president of this club.

David Trauth, age fifty-two, a lawyer from Newark, New Jersey, and **John C. Wehmann**, a civil engineer from Maplewood, New Jersey, were on Flight 258. **Harvey L. Schwamm**, age fifty-three, a banker from the Bronx, New York, was traveling with his wife **Lillian Schwamm**, also age fifty-three. **Jewel Shabell**, age twenty-one, a graduate of Cornell University and resident of Flushing, Queens, New York, was traveling to Nantucket for a vacation, as was **Ira E. Wright**, age thirty, a Yale graduate and industrial designer employed by Hodgman-Bourke, Inc., from New York. He was headed to the island to join his parents, who were already there on vacation.

And there was my mother, **Mrs. Jacqueline A. Young**, age eighteen, and me, **Cindy Lou Young**, age eighteen months.

*

When the weather is bright and sunny, Nantucket lights up with a beauty all its own, a beauty visitors fall in love with when they come to the island. However, any Nantucket native will also describe, in most vivid terms, the familiar sight-limiting "pea soup" fog that rapidly blankets the island when it rolls in from the ocean over the banks of the south shore. There is a saying on Nantucket: "fog happens," and it happens a lot during the month of August, when the weather is usually hot and humid. It happens so commonly that Nantucket is often referred to as "the grey lady, as the island is frequently covered by fog or clouds.

The last official log of the U.S. Weather Bureau at the Nantucket Memorial Airport for that night was recorded at 11:31 PM. It reads: "Sky partially obscured by fog, visibility 1/8 of a mile." The "pea soup" was rolling in off the ocean.

Nantucket's airport is located close to the south shoreline of Surfside Beach. Because of the location it is quickly affected when fog begins to roll in, consuming all familiar landmarks in its path. Islanders who are completely familiar with the landscape often find themselves disoriented in the midst of this thick fog.

At 11:28 PM the FAA Flight Service Station radioed a transmission to Captain Burnham indicating that there was partial obstruction by fog with visibility at a half a mile. Thomas Giffin later recalled this conversation during the official investigation of the accident . He indicated that the communication between himself and First Officer David Carey was proof that there had been an acknowledgment of the first transmission of diminished visibility. The team of investigators had found an omission in the written log – there should have been a timestamp confirming this last transmission, which is why it was noted in the accident report.

In his testimony, Mr. Giffin went into detail about the conversation. He recalled asking the co-pilot, "Did you get that half-mile, Dave?" First Officer Carey replied, "Roger. You know how I love Nantucket." Mr. Giffin radioed back, "Don't plow it up."

Under the FAA regulations, the plane was still permitted to land, the half mile of visibility being within the established guidelines for safe landing. At 11:33 PM, however, five minutes later Flight 258 received another transmission indicating that visibility had now deteriorated to one-eighth of a mile due to the fog.

There was no indication that this message was ever acknowledged by Flight 258, although Mr. Giffin reported that he had heard a "click," and thought that it might have been an indication that the message had been received by the crew on the plane.

Another plane, Northeast Flight 2287, was sitting on the runway, awaiting takeoff. Scheduled to depart after the landing of 258, there was a logged communication between the captain of this plane and ours. Captain Raymond Roy asked Captain Burnham how the weather was over Martha's Vineyard, knowing that our plane had just passed by the island along its flight path into Nantucket. Captain Burnham responded, "Martha's Vineyard should be no problem," and then asked, "How is it down there?" Captain Roy responded, "Not too good."

During the accident investigation, Captain Roy recalled that there was a rapid sequence of radio transmissions indicating visibility at one-half, then one-quarter and finally one-eighth of a mile. He also recalled no acknowledgement from Flight 258 to any of the three transmissions. Airport workers testified that the "click" which was thought to be a response from Flight 258 could just as easily have been generated by the radio equipment of Flight 2287.

The last transmission of the one-eighth of a mile visibility was below acceptable FAA requirements for a safe landing of the aircraft. Nantucket Memorial Airport was not equipped with the necessary technology to support an instrument landing for in-bound planes under fog conditions. It also lacked an airport tower for monitoring takeoffs and landings.

*

At approximately 11:35 PM the plane was observed by those waiting at the airport passing overhead traveling in an east/north -easterly direction. The plane could be seen through breaks in the sea fog that had only grown heavier as it continued to move in off the ocean in layers or waves on the ground. Witnesses indicated that the stars were still visible on and off as they looked upward through the holes in the fog watching the plane pass overhead.

Frances Karttunen, a high school teenager, remembers hear-ing a plane overhead as her family was preparing to go to bed. Her mother asked, "What's wrong with that plane?" Frances heard the plane, but didn't think anything sounded unusual. Be-fore long, the sirens of fire trucks could be heard as they rushed to the airport. The plane had crashed. Frances wonders to this day if her mother had had a premonition that night. That brief conversation in her bedroom is indelibly etched on her memory.

Soon after the plane passed over the airport, a flash was seen behind the fog bank by onlookers. Flight 258 had undershot the runway by 1450 feet, veering off of the center line by some 600 feet.

The plane landed in an uneven clearing and was believed to have crossed Old South Road, shearing its wings on the tops of pine trees as it began to roll rapidly to the left. Approximately 400 feet from the initial ground contact, the left wing of the plane struck the ground. The plane began to disintegrate progressively as it continued to drag its wing an additional 300 feet.

The aircraft then cut through a narrow swath of scrub pines and catapulted into a vertical position. After reaching a position slightly past inverted, the aircraft slammed back into the ground, making simultaneous contact with the nose section and right wing. The center fuselage, however, continued to cartwheel forward, sliding an additional 125 feet, now 1,100 feet from the point of initial ground contact.

As the aircraft finally came to a stop, the center section, which had broken away from the fuselage, was now sitting 90 degrees to the left. Fuel from the shattered wings was hurled into the main wreckage area, igniting immediately into a raging fire. The fire consumed a major portion of the plane's wreckage, along with the bodies of many of the victims of Flight 258. One of those victims was my mother.

Annie and Arnie, who were there waiting to see if we had made it on time to catch this flight, watched as the plane passed over the field heading eastward. Then, they too saw the flash and joined the rush of onlookers toward the scene to assist.

They had no idea if we were on the plane. My grandfather has told me of the desperation he felt as he searched through the rubble trying to save victims, all the while praying that we had not gotten on the doomed flight – but also frantically looking for some evidence that if we had, we had survived. Even as I write this, knowing his memories of this night, I still find myself unable to fully comprehend or imagine the emotional trauma he and Annie had to have experienced.

Annie's recollections of that night remained more vivid through the years. They included haunting memories of running blindly through the fog, through dark, low-lying brush. She could never forget the putrid smell of jet fuel, burning debris, human flesh and hair that night. She would tell me of her panic as she tripped over body parts while she stumbled on through the darkness to the impact area. To say this escalated her alcohol addiction to a greater intensity would be an understatement. Listening to her memories of that evening, I accepted this as the reason for her fragile emotional state.

As he saw the flash through the fog bank, the pilot of Flight 2287, still sitting on the runway, radioed to Mr. Giffin "There's a fire out there".   The airport siren signaling an emergency event was sounded.

In all, twenty-two passengers perished immediately upon impact at the crash site, including the three crew members. Of those that perished, eighteen were burned or charred beyond recognition. Positive identification of the remains would continue into the late hours of the following day. Those who perished on impact included:

- Captain John Burnham
- First Officer and Co-Pilot, David Carey
- Stewardess Arlene Dabek
- Eric Bannister
- Ms. Barbara Dasch

- Mr. Gordon E. Dean
- Mr. Al Foster
- Miss Jeannie Cavin
- Miss Barbara Gillen
- Miss Paula Haken
- Miss Helen B. Hereford
- Mr. William W. Hinckley
- Mr. Edward Kilberg
- Miss Virginia R. Marsh
- Mr. John S. Pearson
- Mr. Sidney Rubenfeld
- Mr. Harvey L. Schwamm
- Miss Jewel Shabell
- Mr. David Trauth
- Mr. John C. Wehmann
- Mr. Ira E. Wright
- Mrs. Jacqueline A. Young

Two passengers would succumb to their injuries the following day, bringing the final death toll to twenty-four victims. They were:

- Mr. Donald S. Breswick
- Miss. Barbara Fienerman

*

Tales of amazing acts of heroism and community spirit emerged from that foggy, dark, smoke-filled night. The grisly task of pulling charred, broken and dismembered bodies from the wreckage awaited the first responders who arrived at the scene. Island residents and visitors courageously and selflessly stepped up to join in this overwhelmingly grim rescue effort.

The dedicated, caring, tough spirit of this small community soon rose from the smoke and ashes of that August night. To this day, there has never been another tragedy similar in magnitude on Nantucket and hopefully, there never will be.

Have mercy on me, O God, have mercy on me, for in you my soul takes refuge. I will take refuge in the shadow of your wings until the disaster has passed.
– Psalm 57:1

The rescue

Many of those who helped with the rescue described the accident as their "worst nightmare." Roger Roche was on duty at the ticket counter that evening. As he awaited the arrival of Flight 258 he, too, noticed the flash at the end of the runway.

Roche and another employee, Pete Giuliani, jumped into a jeep and raced toward the area where they thought the flash had occurred. Raymond Roy, captain of Flight 2287, Jack Kapopolus, a Northeast ramp agent, and the stewardess on Flight 2287 also abandoned their plane and jumped in a vehicle to join the others rushing to the scene.

Phones began ringing at numerous emergency response locations on the island, including the police station, the hospital, the Red Cross emergency center and the island's Coast Guard Station. Soon there were hundreds of volunteers making their way toward the crash site and to the hospital. Sirens broke the silence of the summer night.

Rescuers found themselves embroiled in what became a long, frantic, unforgettable evening, an event that quickly became a legend in Nantucket history, an example of both the island's darkest hour and its finest, with regard to the efficiency and caring efforts of all who were involved in the rescue.

Les Costa and Folmer Stanshigh, both employees at the airport, were among the early responders to the scene. After speaking with the crew on the DC3, they jumped into a jeep and headed toward the end of the runway from which they thought the flash had come. Reaching the runway's end, they broke their way through the scrub pines, making their way further into the wooded section The fog and smoke was by now so thick that they could barely find their way across terrain that under normal conditions was extremely familiar to them. The debris area was now illuminated by the raging fire that was being fed by the spilled jet fuel.

All of the first responders faced the same struggle through darkness and the dense scrub brush. The scope of the horror and destruction awaiting them became apparent to them only as they neared the glowing area.

Awaiting their help in the vicinity of the impact were the few dazed and shocked survivors of the crash. The plane had splintered and smashed into pieces. Trapped in the twisted and burning rubble that had once been the plane's tail section were a few victims, still alive and crying out for help.

Among those victims sat my mother. She had not perished on impact. One survivor remembered her instructions to him to "save the baby."

*

Mary Gore Dean, the wife of one of the victims, Gordon Dean, was waiting at the airport to meet her husband; she too watched as the plane passed overhead. When she saw the flash, she joined the rush toward the point of impact. She later told her daughter, Deborah Gore Dean, about her haunting experience, recalling her frustration with some of the men who were standing hesitant at the edge of the debris area. Worried about moving in closer to aid the victims who were crying out, they feared that more jet fuel might explode with the searing heat. Broken pieces of wreckage and smaller fires throughout the area also blocked access to the site.

Ignoring the chaos and danger about her, Mrs. Dean bravely made her way toward the cries, hoping desperately that one of them might be the voice of her husband. Charred and fractured bodies were strewn about the crash site, mingled among passengers' personal belongings. Many rescuers later described the horrendous experience of trying to decipher human remains from broken wreckage in the murky darkness, bodies twisted in grotesque, contorted positions, many covered with oil or charred and blackened from the fire.

*

Annie and Arnie also made their way through the darkness and debris toward the impact site. Arnie searched frantically for signs of our presence, simultaneously assisting with the rescue of many of the other victims as he went. At one point, he and Annie drove back to the main road to help guide the emergency response vehicles toward the area where they were needed.

In the midst of the dazed and shocked survivors was John Shea, the Wall Street attorney. Mr. Shea had heard a woman screaming from the tail section a few minutes after impact. The words he heard: "Take the baby!"

He told reporters later that he had taken me from my mother's arms and placed me under a pine tree before any of the rescuers had arrived at the site. Once help had arrived, he was unable to remember where he had placed me, or even how far away the tree might be; he had slipped into shock as a result of his own injuries. After rescuing me, he had gone on to help two other women, thinking that one of them might be my mother. Neither were; my mother had remained trapped in the burning wreckage. By the time rescuers reached the site she had perished in the flames.

*

Some of Lita Levine's rescuers included Folmer Stanshigh, Donald Oliver, Oscar Ceely, Jim Cranston, Stanley Conway, Paul Stojak, and my grandfather. There were others. Allen Holdgate, another Nantucket native, assumed the role of coordinating the monumental effort, directing workers where they were needed most. Rescuers worked frantically, battling the heat and flames, spraying their bodies with water from a fire hose to prolong the amount of time that they could remain in the midst of the heat from the burning wreckage. They struggled for hours trying to lift the heavy engine off the trapped woman.

Lita remained conscious the entire time the men worked to free her. She cried out in pain each time the engine moved on top of her; regardless, she kept encouraging them not to stop, crying out continuously that she didn't want to die. Later, she would explain how frightened she was that they would give up.

The rescuers attached a rope to the engine, then tried to gain enough leverage to lift it off of her. At one point they feared that the rope would break, dreading the possibility that the heavy engine would drop back down on Lita, creating even more injury and pain. Through it all, she kept begging them to continue pulling, to save her from her plight.

Not only was Lita in extreme pain, she was also worried that she had lost her sight. Not realizing that the spray from the fire hose was hitting her full in the face, she thought that had been blinded as a result of the crash. The rescuers tried to assure her that her vision and eyes were all right.

One rescuer, Charlie Davis, sat holding Lita's hand, trying to comfort her throughout the ordeal. At one point he leaned back on his other hand to rest, only to place it on the face of a victim who had already been claimed by the crash, also unrecognizable under all the oil that had spilled.

In total, twenty rescuers assisted in the effort to free Lita Levine. Some worked to lift the heavy engine, while others dug frantically beneath her body in an effort to relieve some of the pressure from the engine's weight. A portion of the engine rested on Lita's stomach. She suffered extensive injuries including blistering burns to her head, face and arms, a gash on her ankle from the cowling of the engine, and third degree burns over fifty percent of her body. She was the last victim to be removed from the site.

*

Not far from Lita were more victims. Some had perished; some were still alive. The main section of the plane that split in half tossed many of the bodies out of the cabin. Two men nearby had been thrown clear of the wreckage, but they both had sustained critical head and neck injuries.

Workers remained at the site until the early morning hours. Some continued to dig through the wreckage. Other rescuers worked to clear trees and brush that were blocking the way. Some worked until they had to return to the airport for their regular 7:00 AM work shift.

Police Sergeant Stuart Chadwick was unable to get close to the burning wreckage when he first arrived, so he went back to the road to help direct emergency fire vehicles and ambulances from the main road to the site. The first fire truck arriving at the scene immediately got stuck in a hole in a wooded area just off of the main road. Sergeant Chadwick helped direct other vehicles away from the hole. All the while, he repeated assurances to the dazed survivors that help was coming. John Shea told the sergeant, "There's a baby on board somewhere. I had a baby, but I lost it."

Returning to the crash site, Sergeant Chadwick came upon a woman sitting all alone, repeating over and over that she could not bend over. Her back had been injured. Not burned like many of the others, she had either been thrown clear of the wreckage or had managed to crawl a considerable distance away.

Emergency vehicles quickly began running out of supplies and stretchers. Workers improvised using anything available. Boards found in the back of trucks were used to stabilize and transport the injured. Most of the victims with painful injuries were very quiet, though some managed to encourage and graciously thank those who were working so tirelessly to help them. Some of the injured even apologized for their moans of pain.

The darkness of the crash site was particularly challenging. George Lusk recognized the problem, rushed back to the airport garage and returned to the scene with a truck equipped with a floodlight. Arriving back at the site, however, he was still unable to get the truck close enough to the wreckage to illuminate the area. Mr. Lusk remained at the crash site throughout the night, supporting many of the rescue efforts.

As flames were finally being extinguished, the area started to become affected by a heavy, dense, thick smoke. With more light at the site, the debris surrounding the area also came into view.

Clothing was draped eerily over the surrounding pine trees. Contents from personal luggage were strewn everywhere. Pieces of the plane's cabin were splintered and littered about.

A group of men who were on duty at the Brant Point Lifeboat Station that night were gathered on the recreation deck watching a football game. A call came into the station reporting the crash at the airport. Quickly nine of the Coast Guards on duty at the Brant Point Station, four from the Loran Station and one guard responded to the scene. Their commander, Isadore L. Souza, brought all available equipment with them. In addition to equipment, they brought several first aid kits. One of their trucks had a winch which was quickly put to use in the effort to free Lita Levine.

As they arrived, they met up with other rescue workers who had been actively working to assist the victims. In passing, one worker commented to one of the young Guards, "It is a mess in there. Prepare yourself – they're coming out in pieces." The young man watched as two stretchers passed by him. One of them carried a body, its legs twisted around its torso and head. The second carried the body of another victim. The only recognizable items were a pair of shoes. The young man mustered his resolve and continued on to the area of the worst devastation.

*

In the midst of the chaos, Sgt. Robert Haley, a member of the Massachusetts State Police, was making his way through the debris and darkness when he heard a small noise some distance from the impact area. He walked toward the direction of the sound, and there in a clearing he found me, standing there, whimpering, under a small scrub pine. My clothes had been singed off. I was holding my charred teddy bear. He picked me up, carried me to a cruiser and headed for the hospital. Covered with pine needles, I sat closely by his side in the patrol car as we rushed away from the scene.

*

Most of the men remained at the site until approximately 4:30 AM when it was believed that everyone on board the aircraft had finally been accounted for. Several others stayed on to secure the crash site until daybreak.

There was a great deal of confusion over the number of victims that needed to be extracted from the site, in part because of names not appearing on the passenger list. Because I was a baby, I had not been included in the passenger count at all. It was airline policy not to count infants as passengers because they did not occupy a seat; I was seated on my mother's lap for the entire flight. Her name did not appear because of our "standby" status.

Roland Huyser, one of the responders, was overcome by exhaustion during the height of the rescue. After working with the team to free Lita Levine, he then continued extracting over a dozen more charred bodies of victims. When Charlie Davis came across him during the effort, he noticed a gash on Roland's hand.

Roland told Charlie that he had cut his hand on the leg of a body he was carrying from the site to an ambulance. Charlie took him over to one of the fire engines, found a first aid kit and bandaged Roland's hand. Roland indicated that he was now "feeling woozy." Other rescuers suggested that he step away from the wreckage and smoke to breathe some "fresher air." Heading toward a clearing where there was less smoke, Roland collapsed. Charlie Davis and Officer Bill Burdick carried him out and loaded him into a vehicle to be transported to the hospital, where he was hospitalized along with other victims from the crash. His hand eventually became infected from the cut.

*

It had been a quiet evening at the Nantucket Cottage Hospital. Nurses had settled their patients for the night. The usual hush that typifies the late shift in any hospital prevailed. There were no forewarnings of the pending storm of tragedy that was about to burst upon them.

At about 11:30 PM the telephone rang. Marion "Goldie" Howes, the night nursing supervisor, answered the call. The caller reported "a plane has just crashed at the airport." No further details were given; the connection was broken. Mrs. Howes immediately placed a call to Leroy True, administrator of the hospital and repeated the message that she had just received. He told her that he was on his way, jumped into his car and headed toward the hospital.

The staff on duty started wondering among themselves if the crash involved a small private plane, or one of the larger "air ships" Northeast Airlines was now regularly landing at the airport. Unsure of the potential size or gravity of the situation, the decision was made to put the facility's emergency disaster plan into effect immediately.

The disaster plan was simple, flexible and efficient; it involved first contacting all of the physicians on staff. Those contacted that evening included Dr. Ernest Menges, Dr. Wylie Collins, Dr. Paul Cassidy, Dr. Charles Sziklas and Dr. Ralph Harvey.

Mr. True notified the hospital's board of trustees. Those that he contacted included Alexander Craig, Charles Snow, Raymond Folkrod, Syd Conway, Jules Thebaud, Lewis Greenleaf, Jr., Mrs. C. Conyngham Gifford, Mrs. J. Winston Fowlkes and Mrs. Hal B. Armstrong. The trustees quickly grasped the potential for the hospital to be inundated by distraught family members, reporters, curious onlookers, volunteers and other non-essential personnel in the wake of the disaster. They too joined the staff at the hospital to help manage the expected flood of people.

Nurse Howes continued working the disaster plan, making phone calls to other nursing staff supervisors, including Irene Chase and Mrs. George Hadden. The nursing staff supervisors in turn called other staff members informing them of the need at the hospital. Nurses who were known to be vacationing on the island were also contacted. Ambulance drivers Gilbert Wyer and Jack Driscoll were already busy at the airport helping with the rescue.

Next, the anesthesiologist was contacted, Dr. Leroy Vandam. He was asked to report to the hospital along with other technicians, Arthur B. Desrocher, Joan Kaufman and Ellen Docca. The head nurse of the operating room, Mrs. James Nelson, my aunt, was also summoned to the hospital. Even the cook was asked to return to the kitchen in anticipation of a long night ahead. Staff poured into the hospital. When Victor Brown and Priscilla Johnson arrived at the hospital, they seated themselves at the hospital switchboard in anticipation of an extremely busy night. The extent of the disaster was still unknown and some of the workers had even begun to question if the initial phone call had been a hoax.

Almost immediately, thirty-eight of the hospital staff of forty-four were on site, gearing up for the expected deluge of victims. Fifteen additional volunteers, comprised of visiting medical students, doctors and nurses, also appeared within minutes of being contacted. All those who arrived wondered what would be in store for them during the next few hours. Staff continued to arrive for several hours after the emergency plan had gone into effect.

*

The sounds of racing car engines, screeching tires and sirens announced the arrival of the first crash victims, who were unloaded from ambulances, station wagons, jeeps and trucks. As they arrived, the staff assembled themselves into several teams to receive and treat the patients for their injuries. These teams worked diligently throughout the evening, comforting patients as they worked to assess injuries and save lives. Victims sometimes came through the door two at a time. There were still enough teams staffed, ready and waiting. Each victim was immediately evaluated by one of the attending physicians.

Ethel Mooney, a nurse who worked at the hospital, was hospitalized herself that evening, having been prepped for a routine surgery that was scheduled for the next morning. As she lay quietly in her room, she heard the burst of activity and the wail of sirens rushing in. She got up out of her bed and joined her fellow nurses, helping to dispense medication and aid to the victims coming through the doors. She was still dressed in her own hospital gown as she too worked to save lives.

Upstairs in the patient area, nurses began pulling extra beds out of the storage rooms, continuing to make preparations for the many patients they anticipated would need to be hospitalized. That expectation was eventually darkened by the grim realization that most of the victims were being moved directly to a make-shift morgue.

Six additional beds were moved into the men's ward, where one patient, Albert Huyser, was recuperating. He moved himself down the hall to a small private room to free up space for the expected incoming patients. Later in the evening he was joined in the private room by his brother, Roland, the man who had collapsed during the rescue from exhaustion.

The hospital facility was new, barely a year old. There were still many beds stored at the old hospital building located near the center of town. A nurse headed over to the old hospital building to get more beds. In her haste to gather the extra beds, she left a headboard sitting on the sidewalk in front of the old facility.

A call went out to Charles Flanagan, the head of the Red Cross. He was asked to help provide additional stretchers in anticipation that they would be needed to unload the victims as they arrived from the crash scene and also to transport them off the island to mainland hospitals for further treatment.

Everyone was now working feverishly to stock the supplies that would be needed for whatever types of trauma the night would bring. Women began preparing coffee and sandwiches for the family members who they expected to start arriving momentarily.

The dead were taken to a small room set up as a temporary morgue down the hallway from the emergency room area. This room filled quickly; soon the hallway became the overflow area for the morgue.

Some patients were sent directly to the patient area upstairs, where they were met by teams in each room for evaluation and treatment. The emergency accident room was quickly filled to capacity. Some of the victims arrived with their clothing melted to their bodies. Nurses carefully worked around this as best they could, in an effort to dress and treat the wounds and burns. Shots of morphine and tetanus vaccine became the routine order for many victims throughout the night. Everyone worked to address whatever injuries presented themselves, but this was a small hospital, not nearly prepared for devastation on the scale of this disaster.

Minor operations were performed at the patients' bedsides. Volunteer medical students followed physicians, assisting with minor treatments and sutures throughout the night. Amidst this overwhelming burst of activity, there was comforting quiet as each staff member went into an almost "automated mode of operation." Everyone confidently worked in unison. No one had the time to wonder "what should I do?"

The victims who were most seriously injured and in pain beyond description sensed the competency among the workers that surrounded them. There were no screams, no panic or hysteria. The patients trusted in the care that they were being provided. Lita Levine was in considerable pain and discomfort, but she kept trying to ask the hospital workers to contact her mother, knowing she would be upset not to have heard from her daughter as promised. It distressed her that her mother did not know her whereabouts or about the grave situation she now found herself in.

Downstairs in the reception area, a steady stream of people began filing in through the entrance, asking "what can I do to help?" and "do you have any information on the dead or survivors?" One woman came in looking for her husband who she believed had been on the plane. Others arrived behind her, asking if they could donate blood or help serve food to those now occupying the waiting area. A well meaning local man came through the door with a basket containing six bottles of whiskey. He was disappointed the next day when he learned that they had not been used.

The switchboard quickly lit up with a continuous stream of incoming calls. The list of survivors was revised several times during the first few hours. It was a short list.

While many in the waiting area were doing their best to support and comfort each other, some family members began to grow frantic waiting to learn the status of their loved ones. Some tried sneaking up to the second floor to see if they could recognize any of the patients that had been admitted to rooms. The hospital staff reacted quickly, posting guards at the second floor elevator and stairwell.

Nurses Hazel Kingsley and my aunt, Louise Nelson, were working with the police and other volunteers in the morgue area, attempting to identify the charred bodies of victims. Family members were brought down the hallway to the morgue two at a time, accompanied by either a priest or a minister, to try to identify loved ones as well.

As victims were identified, Dr. Ernest Menges, the medical examiner, declared the time of death for each victim's death certificate. The family members, like the injured victims upstairs, faced this grim task with an incredible amount of courage and grace.

*

During one of her trips back to the waiting area, my aunt noticed Annie and Arnie seated in the waiting room with the other family members waiting for any news. Making her way over to them, she asked them why they were there. My grandfather explained that he was unsure if my mother and I had been on the plane. She quickly arranged for him to go to the morgue area to see if he could recognize anything familiar on any of the corpses that had already been transferred there.

Arnie decided it best to go to the morgue area without Annie, anticipating the grotesque scene that awaited him. He did not want Annie to be left with lasting memories of the charred and blackened bodies of her daughter or granddaughter. He headed down the hallway with my aunt and a minister.

Entering the make-shift morgue, Arnie passed by several bodies, unable to recognize anything identifiable. Then as he started to pass another body, he spotted a familiar item that he had so hoped he wouldn't see: his daughter's wedding band. He broke down and wept as he realized his worst fear was now a reality. He made the identification for the medical team standing at his side. This body was his daughter Jackie, my mother. The doctor declared the time of death.

Arnie looked at the remaining corpses; there wasn't one small enough to be me among them. Then he turned and walked slowly back toward the waiting area, dreading the news he now had to deliver to his wife. He took as long as he could, trying to prepare himself to break the news to Annie in the gentlest way possible, knowing all too well the heartbreak that would follow.

Glancing up the hallway as he approached the waiting room, his eyes met Annie's frightened gaze. Annie took in his glistening blue eyes and the streaks of tears on his cheeks. There was no need for words between them. Hope vanished and the answers to her question were now painfully clear. "Oh, God, no!" Annie cried out as she collapsed in the arms of the strangers around her.

Annie and Arnie were ushered by nurses to a room where they could share their grief privately. A short time later, another nurse entered the room to tell told them that I was upstairs, checked in but apparently uninjured. She encouraged them to go up and see for themselves, as soon as they felt they were ready. The staff hoped that this small bit of good news would lighten the burden of grief and sadness that had overwhelmed the couple.

Sometime later, Annie and Arnie made their way up to the second floor, where I lay quietly sleeping. Again, they were flooded with emotions, many of them mixed. I had survived, but my mother, their daughter, had not. What was in store for them next? What would tomorrow bring?

Emotionally drained and exhausted, they returned to the small private room to sit quietly with their grief and wait for the morning light which was only hours away.

Neither this man nor his parents sinned" said Jesus, "but this happened so that the work of God might be displayed in his life. – John 9:3

The day after

Finally the morning light broke through the darkness of the chaotic night. The hospital was now taxed beyond capacity. Of the twelve survivors, seven were in critical condition from burns and fractures and required treatment beyond the capabilities of the small island facility. The fog still presented problems as hospital staff realized that these seven needed to be transported immediately to Boston for further treatment.

The hospital contacted the army for help transporting these victims. Otis Air Force base responded with a C54 Air Rescue plane equipped with a team of doctors, Corpsmen and medical equipment.

Among the survivors who were transported were **Marguerite Boos**, who was transferred to Boston's burn center, and **Mr. Edward Fennell**, who was listed in good condition shortly after he arrived in Boston. **Donald A. Breswick**, who was listed in critical condition, he succumbed to his injuries soon after arrival at the hospital in Boston.

Lillian Schwamm, wife of the deceased banker **Harvey Schwamm** was critically injured with extensive burns and cuts. **Lita Levine**, the young artist who had been trapped under the plane's engine was also transferred to the burn unit at Mass General. **Marylin Pearson**, wife of deceased crash victim **John Pearson**, was transferred in critical condition with a fractured hip and shoulder. **John P. O'Dell** was transferred after x-rays showed possible fractured vertebrae of the neck and a broken collar bone.

As the plane lifted off from Nantucket, it was evident that seven lives hung in the balance. Air Rescue's commission was to get these victims to the appropriate medical facility in Boston as quickly as possible.

Remaining in Nantucket Cottage Hospital were passengers **Paul Kozinn**, who suffered bruising, black eyes and a gash on the back of his head and **Joyce Bell**. Both were listed in good condition. Bell remained at the Nantucket hospital with her eyes blackened, scratches on her body and her head in a nine pound traction sling. **John J.B. Shea** was also listed in good condition; he suffered bruising lacerations and back injuries.

I, **Cindy Lou Young**, was listed in good condition as well. My only injury was the small scratch on my chin.

*

As the medical evacuation plane departed from Nantucket, other flights began arriving filled with additional medical personnel, reporters and family members. The island was soon swamped with people.

Meanwhile, teams of doctors and dentists worked together to review individual medical and dental records in a continuing effort to identify those victims burned beyond recognition. Representatives of Northeast Airlines on the island made arrangements for the surviving family members. The day was filled with events all set into motion as a result of the crash.

Northeast Airlines dispatched a plane to Nantucket with doctors, nurses and company personnel. The crash made headlines on Nantucket, Cape Cod, New Bedford, Boston and New York. *Life Magazine* covered the crash in their August edition.

Sometime during the early morning hours, my father received a phone call from the hospital notifying him of my mother's death and my condition. He and his mother quickly made arrangements to travel to Nantucket.

Meanwhile, Annie and Arnie had filed an emergency petition with the court to take custody of me. A court order was immediately issued granting them temporary custody. As soon as the hospital workers learned about my father's intentions to travel to Nantucket, they informed Annie and Arnie. This day began the deep divide between my families, one that would last forever.

*

One thing that I have learned through my own struggles with addiction and alcoholism is that nobody can cook up resentment in his or her head better than someone affected by this disease. This, I believe, is what took over in the midst of the turmoil of emotions that were consuming Annie and Arnie in the immediate aftermath of the crash. The same people who had once wanted my father to come to the island for blue fishing and had sized him up for hand knitted sweaters, instantaneously decided that he was "the enemy" – the cause of their daughter's death, the sole person responsible for this disaster in their lives.

Islanders are very loyal and protective of "their own." Community members soon closed rank, rallying to establish a protective shield around Annie, Arnie and myself. They were fully prepared to keep anyone from removing me from the island and the family that they knew. Island hearts turned cold. When my father and paternal grandmother arrived on Nantucket they were greeted by a community that no longer welcomed their presence. They were not comforted in their loss as so many others had been; rather, they were shunned and rejected. The situation was truly hostile.

Realizing the futility of the situation, they returned to Ohio without me and began the process of obtaining their own legal counsel to represent their interests. They made no further effort to challenge or change the temporary custody arrangement that had been issued in the midst of the disaster. They, too, needed to process the situation and the impact of this event upon their own lives.

*

In Washington, D.C., the United States Senate held an unusual Saturday session, solely for the purpose of paying tribute to Gordon Dean. The session was led by President Eisenhower and Dean's successor at the Atomic Energy Commission, Lewis Strauss. Both, the president and Mr. Strauss began the session expressing their shock at the sudden death of their colleague. In the later part of 1943, Dean had joined the United States Navy as an intelligence officer. Later, in 1945, he was selected to assist in the prosecution of Nazi war criminals, assisting U.S. Attorney General Robert H. Jackson in drafting the plan for the International Military Tribunal. In 1949, Dean had been appointed to the Atomic Energy Commission. He and Mr. Strauss together led the efforts to reverse the commission's recommendation against the development of the hydrogen bomb. Dean led the commission over the next several years, a challenging period of vast expansion of nuclear facilities, materials and weapons, serving in this capacity until the completion of his term in 1953.

Others who paid tribute to Gordon Dean that day included Senators Henry M. Jackson from Washington, Estes Kefauver and Albert Gore, Sr., all Democrats from Tennessee; Senator Wayne Morse, a Democrat from Oregon; Senator Michael Monroney, a Democrat, from Oklahoma; and Senators Francis Case from New Jersey and John S. Cooper from Kentucky.

Senator Gore, as mentioned a cousin of Dean's wife, Mary Gore Dean, led the tribute acknowledging that this tragic death called attention to the quality of Mr. Dean's citizenship and service to this country.

Senator Jackson noted that only the fateful previous night, August 15th, some of the recommendations made by Gordon

Dean as a member of the Atomic Energy commission had been approved by the State Senate. He noted that the nation had suffered a great loss with Mr. Dean's passing in the prime of his life with so many fruitful years ahead of him.

Senator Cooper expressed that "when the history of the development of atomic energy is known, the contributions of Gordon Dean toward its development will be regarded as one of the great contributions in that field."

*

At home in the Bronx that Saturday morning were Nathan and Beatrice Levine, passenger Lita Levine's parents. The couple had not really known where Lita was headed for her vacation; they only understood it was somewhere near Boston and that she had promised to call her once she reached her destination.

Throughout that evening they grew more and more concerned, not receiving the promised phone call. At about 3 AM Beatrice woke up, seriously worried. She continued to toss and turn until the early morning hours.

As Nathan awoke the next morning and went out, as usual, to walk the dog, he noticed the neighbors gathered looking at the headlines in the morning paper about a plane crash on Nantucket Island. Noting their curious looks, he gathered his paper and went back into the house.

Inside, Nathan asked Beatrice once again where Lita had been headed. Beatrice again told him all she knew, somewhere near Boston. A few minutes later, the phone rang. A man on the other end of the line informed Nathan and Beatrice that there had been a plane crash and that Lita had been on board. He kept repeating that Lita was in fair condition.

As Beatrice hung up the phone, neighbors who had already read the headlines began to gather to comfort the couple and help them in any way they could. Beatrice asked Nathan to call her brother-in-law in Mount Vernon, to see if he could assist them in making arrangements to go to Nantucket. They wanted to get to their daughter's side as quickly as possible

*

As they were making arrangements, Lita and the other six survivors were arriving in Boston. A police escort and ambulances met the plane at Logan Airport. The call had gone out to all emergency medical personnel in the area that help was needed to assist in saving the lives of the victims of this crash, all of whom were badly burned.

It was a typical hot mid-summer day in the city. People woke up to the non-stop commotion of sirens blaring through the city streets, transporting the victims from the airport. Within eight hours of the crash, emergency personnel and all available life-saving equipment had been assembled and readied. At the burn unit at Massachusetts General Hospital, teams of specialists that included a doctor, two nurses, a hospital aide, house staff and a medical student were assigned to treatment rooms making preparations to receive the victims as they arrived. The burn unit had been established to respond to extreme burn cases like those which had been experienced after the 1942 Coconut Grove nightclub fire. That fire took the lives of 492 people; hundreds more were seriously injured.

Stretchers began blasting through the emergency room doors. On one of the stretchers lay Lita Levine, covered in blankets. Her hands, severely burned, looked as though she had white gloves on. The doctors began the lengthy process of saving the young woman's life.

Lita's story is a compelling example of the human spirit and the will to survive. Drifting in and out of consciousness, she repeatedly mumbled the request to contact her parents. She was still terribly frightened that she had been blinded in the accident; both eyes were bandaged to cover the burns on her face. One of the doctors removed the bandage over her left eye momentarily to reassure her that she was still able to see. Her worst fear quelled, Lita drifted back into unconsciousness.

Her doctor, Dr. George Nardy, made a full assessment of Lita's condition. Clearly, she was badly burned. He was particularly concerned about his ability to salvage her hands. Dr. Nardy once stated his belief that burn injuries are the most traumatic injuries that man can endure, recognizing that this type of trauma challenges every aspect of medical care.

Around noontime, Lita's parents arrived at the hospital. They sat in a waiting room with other families of the victims, trying to console each other. Mrs. Levine recalled an overwhelming sadness that enveloped the room as they all realized just how young the victims were.

Nathan and Beatrice soon got their first glimpse of Lita. Her head was bandaged, as were her hands and legs. She awoke momentarily and said, "Hi, Mommy. Glad you're here." She then asked her mother how she and her father had gotten there "On a duck's back, honey," Beatrice answered, not wanting to mention an airplane. Lita asked her mother not to leave her. There was no possibility of that ever happening; Beatrice was there for the duration. Lita asked her mother where she and Nathan would stay. Beatrice responded on the floor if need be.

As the day ended, Lita drifted in and out of sleep. She fought the medication, fearful of being left alone in the darkness. The lights in her room remained on day and night for many days. She finally gave in and fell into a restful state. It had been one long, life-changing day for her and the ten other survivors of Flight 258. Lita remained in her hospital bed in room 601, unable to move for months.

*

Back on Nantucket, the airport and hospital workers had all gathered at the local bar in town to unwind after their hectic night and day. One of the nurses mentioned how sad it was that all of my clothing had been lost or burned. A pilot sitting at the bar suggested that everyone put money in a jar every time they swore.

He suggested that the proceeds could be used to buy me some new clothes. The language immediately grew atrocious in the bar and shortly after I received a brand new wardrobe.

PICTURES

1950's Northeast Airlines Vacation poster

Picture of a Northeast Convair, printed with permission of Bill Armstrong

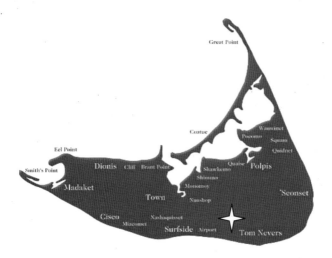

A map of Nantucket Island, the star indicates the location of the airport.

Captain John Burnham, Northeast Airlines Flight 258

An aerial view of runway 24 . The arrow in the bottom right hand corner of the image points to the location of the crash.

Smoldering wreckage at the crash site, August 15, 1958.

Rescue workers trying to push the engine off of Lita Levine.

Rescuers surveying the engine that trapped Lita Levine.

Rescue workers at the scene.

Rescue worker,
Charlie Davis

My angel, John J. "Jack" B. Shea, he responded
to my mother's last request to "save the baby".

Rescue worker,
Donald Oliver

Rescue worker,
Allen Holdgate

Hospital Administrator, Leroy True

My mother, Jacqueline Anne (Duce) Young

This picture included a caption that my mother, Jacqueline Anne (Duce) Young , was found in the wreckage in the area of the circle. My grandfather, Arnie, is searching through the wreckage on the right in the plaid shirt.

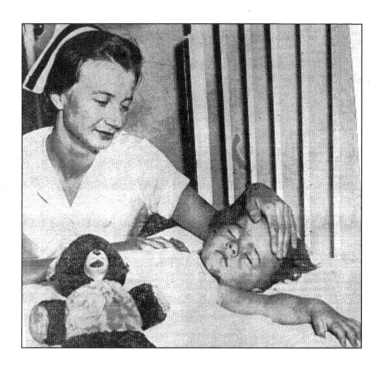

This picture, taken just after I was admitted to Nantucket Cottage Hospital, appeared in the August 1958 issue of *Life Magazine*. Nurse Beverly is at my side watching over me.

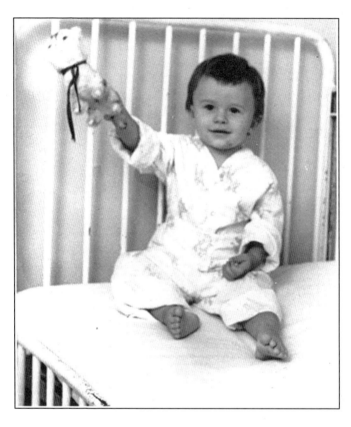

The next day I was happy to entertain all of the reporters and visitors that came into my room. My only injury was the small scratch on my chin.

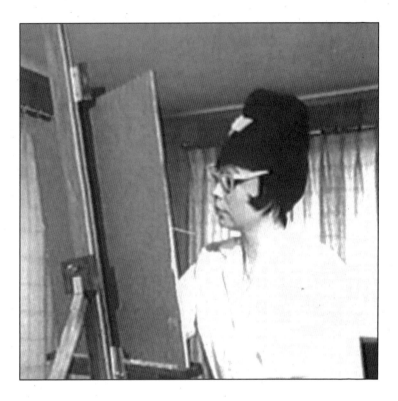

Lita Levine works on one of her paintings after her recovery.

Beatrice and Nathan Levine, Lita's parents, during an interview for the *Dupont Show of the Week, "Patient in Room 601",* which aired in 1964.

Happiness is like a butterfly. The more you chase it, the more it will elude you. But if you turn your attention to other things, it comes softly and sits on your shoulder.
– Unknown

Life goes on

The days, weeks and months that followed were filled with challenges for all of the survivors, their families and the surviving family members of the victims who perished.

Mrs. Dean struggled for many years to return to a life that was not plagued with haunting memories of the crash that night. She moved in with Al Gore's family, in Tennessee, who helped her care for her two young children while she worked her way through her own grieving process.

Survivors Paul Kozinn and John Shea also completed their own recoveries. Each was left with haunting memories of the accident and the curious question of why their lives were spared.

Lita Levine's mother and father rented a room at a hotel in Boston when they first arrived in town. A representative from Northeast Airlines made arrangements for them to move to the Hampshire House, which provided more comfortable accommodations. It was clear that the couple would remain in Boston for a lengthy time while their daughter recuperated.

Every morning at 6 AM, Nathan walked over to the hospital and went directly up to his daughter's room. He was usually there when the doctors made their rounds and the nurses changed his daughter's dressings. Three nurses were needed to move Lita to a new position in her bed.

There were times when Nathan had all he could do to hold back his tears. When he could not, he would quickly take himself to the "gents" room right outside Lita's door and cry. At one point, after the bandages had just been changed, he was so overcome with emotion that he was unable to exit the room. He buried his face on Lita's bed. Lita asked him, "Daddy, are you crying?" Nathan denied it. He said he was sweating because it was hot in the room.

Lita's mother could not be in the room when Lita's bandages were being changed. Even though she would retreat to the floor below, she could still hear Lita's screams. Through the pain and the heartbreak, Beatrice stayed strong. She never once let Lita see her tears.

On August 21st, 1958, Lita made her first trip to the operating room. Forty-one more trips would follow as Lita traveled her path to recovery. In the operating room the doctors noted that Lita's hands were necrotic; most of the tissue was dead. The only viable tissue that remained was on her palms, and even that was questionable.

On this first day, the team of doctors completed surgery to remove the charred stumps that once were Lita's fingers. As Lita awoke from the surgery, she was not told that her fingers had been removed, fearing the news would send her further into depression.

At one point a nurse made a callous comment to Lita's mother. "If she were my child, I would let her die," she said. Beatrice responded to the remark with a simple affirmation. "Well, she is not your child. She is mine and she is going to live!"

Beatrice was one tough cookie from the Bronx and quickly got on the nerves of many of the hospital staff as she looked out for her daughter, making sure that the nurses stayed "on top of things." Nurses threatened Dr. Nardy with ultimatums like "either she goes, or we go."

Dr. Nardy took Beatrice aside one day and tried to help her understand the stress she was creating among the staff. He gently told her that he would hate to have to keep her away. Beatrice answered bluntly, "You just try to keep me away!"

Lita was a prisoner of her injuries, her days filled with fear and excruciating pain. A young woman who had once taken great pride in her own independence, she was now forced to be reliant on others in every aspect of life, unable to change herself, feed herself, or even simply move.

At one point, Dr. Nardy made a note in Lita's medical record to the staff psychiatrist asking, "Do you have anything to offer the girl that has lost her hands?" She had to be told.

The day finally arrived when Dr. Nardy told Lita she had lost her fingers. In a small voice, she asked him, "All of them?" Yes, he told her. "Even the thumbs?" Dr. Nardy nodded, yes. Lita wept at the news.

Soon after, Lita was told, her bandages would be removed, allowing her to see what was left of her hands for herself. Lita hesitated, asking Beatrice to look first. She told her mother, "I'll look, if you do."

Lita's mother remembered this as something she didn't think she was capable of doing. But she mustered up all the strength she had in her and watched, unflinching, as the bandages were removed. When she saw the stumps that were once the little hands she had so admired when her daughter was born, she picked them up without hesitation, kissed them and told her daughter, "They're still beautiful"!

Lita's biggest grievance was against God. She often asked her mother, "How could there be a God who would have allowed this to happen? Why did he let this happen to me? What did I ever do to deserve this?" Unable to resolve those feelings, she could not accept that what had happened to her was not God's doing.

Eventually, Lita sank into a severely depressed state. She no longer had the will to live, and told her mother so every time Beatrice was in the room. After a number of days of seeing her daughter in this state, Beatrice became desperate to wake her daughter up to the life that she knew was still worth living.

One afternoon she went into Lita's room. She had with her a small bottle of aspirin. Lita would have no way of knowing what kind of pills were in the bottle; the bottle had no label. Once again, Lita expressed her desire to die, and again Beatrice patiently explained to Lita that God did not do this to her. Her mother also told her that if she wanted to die, then she needed to recognize that there were two other people who loved life, and loved her, who would also die with her. She asked Lita, "How will you stand before God and explain to him that you are responsible for the lives of two other people who wanted to live?" Lita did not respond.

Beatrice took the small bottle of pills from her purse and placed them on the bed tray in front of Lita where Lita could see them. "Okay, then," she said. "Let's do it. Stop wasting my time here. It's too bad that Daddy isn't with us – he will have to figure out a way to do this on his own."

Beatrice gave Lita a moment to think, then said, "You decide. If you want to do this, we can do it together. Right here, right now. If not, then we can work through this together. The decision is yours."

Beatrice went over to a chair, dropped herself into it and sat in the heavy silence of the room. Ten minutes felt like days passing. Finally, Lita asked her mother, "What about my hair?" Beatrice replied, "You'll see, honey, we'll get it." Lita asked, "What about my clothes?" Beatrice again replied, "You'll see, honey, we'll get it". Lita asked, "What about high heels?" Again her mother replied, "You'll see, honey, we'll get it."

It seemed to Beatrice that she repeated that same response a million times as her daughter questioned how she would return to the life that she once knew.

After a long silence, Lita finally said, "All right Mother, I'll live." Beatrice picked up the bottle of pills and flung them out the window. This day marked a critical turning point in Lita's recovery. The work, however, was just beginning.

Several of the people involved with Lita's rescue got to meet with her again after she gained a full recovery. One of the people who met her was Bob Mooney. He remembered her wearing a wig with sun glasses. She also wore gloves to cover her hands.

*

My early childhood years were filled with Annie and Arnie fighting over who would be granted custodial privileges as my guardian. My family in Ohio would make many trips to the island over the period of the next few years in their attempt to obtain custodial rights. The court sessions were filled with allegations about Annie and Arnie's struggle with alcoholism. The battle over who would gain custody of me lasted for years.

Judge Sullivan made the final decision to grant guardianship to my great-grandmother, Nana Lamens, a decision that meant I would remain on Nantucket. She was assigned the overall responsibility and oversight of my care even though I was placed in the physical custody of Annie and Arnie.

Defeated, my Grandmother Young and my father went back to Ohio. I would not see or hear from them again for a very long time, until I was in my late teens.

*

Growing up on Nantucket is a unique experience. Although my childhood years were chaotic and plagued by the unpredictable and uncontrollable affects of the disease of alcoholism, living on the small island was an experience that one cannot fully appreciate unless they actually live it. The community was like an extended family. We all looked after each other.

Not every day was chaotic. Our family shared some wonderful times through those years. Just as the nation was going through the changes of the Sixties, so was Nantucket – only just a bit slower.

Summers were filled with long picnics at the beach. We would load up the car with food and family and go have a day at the beach, eating, swimming, enjoying the sun and the company. Winters were long, desolate and quiet.

Winter, however, was one of my favorite times to spend time by the ocean. I used to love to walk the beach during the winter, or drive down to look at how furious the sea had become with each approaching storm. Spring and fall were filled with excitement and anticipation of the upcoming busy summer season or the quiet winter season.

The first real memory I can recall is that of a sunny afternoon when we were living on the corner of Union and Washington Street. I was playing in the living room. Arnie and Annie were in the kitchen. They had started drinking earlier that day. They began to argue and it soon got loud and ugly. Annie picked up a bottle on the table and smashed it over Arnie's head. I had been peeking through the doorway, frightened by the shouting. I was about three or four years old.

I crept to the front porch of the house and ever so quietly snuck out the front door. As soon as my feet hit the sidewalk I started to run as fast as I could to the place that I knew I would be safe and far away from the fighting.

I ran to Nana Lamens. She lived a few blocks down Union Street in a cottage behind the American Legion. I had no shoes on my feet, but that did not slow me down. When I got to her house, Nana was shocked to find me crying on her doorstep. She quickly brought me in, sat me down and asked me what had happened.

I don't remember what I told her – I only remember her picking up that heavy black telephone and calling Annie. She started speaking in Dutch, very fast and very angrily. I know now that she must have really laid into Annie, though I couldn't really understand that at the time.

I stayed at Nana's that night. She fed me supper and soon after that tucked me in on the other side of her own big metal bed. She read me a short story and off to sleep I went for the night. I was calm now, I was safe.

I stayed at Nana's for a few days. I guess during that time, Annie and Arnie sobered up and made their renewed commitment to getting "back on the wagon." They must have, at some point, convinced Nana that they were serious about it. I don't think she would have let them take me back if that had not been the case.

I went back to the house with Annie and Arnie, and for a while things got better. Arnie was a beer drinker, but he also liked to dabble in the liquor every once in a while, picking up Partner's Choice at the liquor store when it was time to "have a little fun." For Annie, it was vodka and pills.

I loved being at Nana's house. Annie and Arnie were both trying hard to hold down regular jobs. Annie was working for Mrs. Anderson as a secretary in town. When Annie and Arnie worked, I went to Nana's house, where I spent my days making mud cakes and pies in the garden, or helping to feed the clothes into the wringer on the washing machine. My favorite days were the ones we spent baking cookies, especially the frosted ones that I got to help decorate. Nana was tough, but we both loved our time together.

*

Then one day on May 9th, 1964, my world took another critical turn. Nana died.

It was on the day of Evan Young's birthday party. I had come home from the party, dreading the cod liver oil I had come to expect, to find Annie crying. She told me that Nana had gone to the hospital earlier that day in an ambulance and that she had died. She had something wrong with her heart, she told me, an angina attack.

After Nana died there was one more court hearing. Judge Poland granted permanent custody of me to Annie and Arnie. Unbeknownst to me, my grandparents were also granted the ability to make monthly withdrawals from a trust fund that had been established for me from the insurance monies from my mother's accidental death.

The account amounted to about $18,000.00 at the time, Annie and Arnie often said that this money was the only reason that my father had any interest in taking custody of me. When we later met for the first time, he denied this as "hogwash." It definitely created a great deal of animosity between all of my family. By the time I was old enough to know about it, there was nothing left. As the saying goes, money is the root of all evil. It certainly brought out the worst in my family.

*

I attended elementary school at the Academy Hill School. Somewhere between first and second grade, I learned that Annie and Arnie were not my Mommy and Daddy, as I had been led to believe. One day on the playground, one of the girls in my class told me that she knew something I didn't know. She then told me the whole story, about the crash and that my mommy and daddy were really my grandparents.

At first I was angry at the girl for telling me this. I didn't want to believe it. I told her she was lying. Then a few more kids who were nearby all started saying that they, too, knew about the crash and that it was true. I was shocked, not knowing what to believe. I couldn't wait to get home that day and find out what Annie and Arnie had to say about this.

I went home after school and asked Annie if what I had been told was true. She got an odd look on her face and then sat me down, telling me as much as she thought I could handle. When I think back on this, I realize she treated me like I was fragile and sure to fall apart at this news. I don't think I reacted badly; I did feel a bit confused by it though.

Annie, I learned later, had struggled early on with the reality of taking on the parental role for me after the crash. She was particularly uncomfortable when I began to call her "Mama," so uncomfortable that she went to her doctor asking what she should do. He told her that this was normal, that I was bonding with the only mother figure in my life. He encouraged her to act "normal" and not discourage me from referring to her as Mommy or Mama, and so Annie did, even though she continued to feel a degree of awkwardness with the situation.

*

When I went back to school I started writing "Jackie" on all of my papers, which really freaked out the teacher, Mrs. Marble.

She quickly called Annie in for a conference on how to handle this situation. The teachers were a bit concerned about my class work anyway. I was tired most of the time and I also wasn't participating like they thought I should be.

Annie promised to keep an eye out and do whatever she could to help the situation improve. They left it alone, hoping that if they didn't make a big deal out of it, I would forget about it, which eventually I did. I stopped the odd behavior – at least that odd behavior. I was the kind of kid who was always doing something – never a dull moment, that was for sure.

Most of my early childhood was consumed with just trying to survive in a world that for the most part was upside down most of the time. Christmas often came with the news that "Christmas was going to be tight this year," the years when Dad (which was what I continued to call Arnie) didn't have a job, or Mom (Annie) was sick and in the hospital again. Sometimes I lived for months at a time at Aunt Louise's house, when Annie checked into a detox facility trying to get sober.

*

One of my best friends during those early years was Margaret Mooney. We were for the most part inseparable. Oh, the things that we used to do to occupy our time! Margaret's family had two homes, one for the winter and one for the summer. When school was out and the tourists were starting to arrive, they would move out to Madaket to their summer home. Margaret's house was another safe haven through my adolescent years. I remember I would call during those boring holidays to see how soon we could get together and do something fun.

During the winters, Margaret and I would spend our time ice skating on Mill Pond, Mioxies Pond (pronounced "Moxies"), and at the flooded tennis courts in town. If the ice wasn't good for skating and there was too much snow, we would spend our days sledding at Dead Horse Valley or some of the other good spots around town. There was solitude in abundance, but it was balanced by more than enough fun. Even in the depths of winter, the long cold days of February, we found plenty of things to do.

Sometimes those things led to a bit of trouble along the way. One time the harbor was frozen solid with ice; no boats could make it in or out unless an ice cutter had made it over to break a channel. One day during this cold snap, as we were walking down by the wharf, Margaret and I ventured out onto the harbor ice. Soon we found ourselves being chased by the local police, who were concerned for our safety more than anything else. Harbor ice varies in thickness and is very dangerous to walk on, which we didn't know. Our minds weren't on ice, however; we assumed the police were following us because they knew about the crank phone calls we had been making down at the phone booths.

One winter day that I clearly remember, I had spent the night at Margaret's house. We had big plans for skating the next day at the tennis courts, which in winter were flooded and made a great skating rink. Margaret lived within walking distance of town and the tennis courts. We woke up that morning to find that Margaret had developed big red spots around her eyes. She knew in an instant they were hives; she'd had them a few times already, usually due to food allergies.

We knew that if her mom found out that she had hives, our plans for the day would be ruined. I told Margaret that I had a cure. I ran downstairs, found a cucumber in the fridge, sliced it up and went back upstairs to Margaret. "Here," I said quickly. "Put these on your eyes and relax." To the surprise of both of us, the redness did appear to improve. We dressed quickly and Margaret put on her skates, the blades covered, so that she would have them on already in case her feet started to swell, another problem with her hives. We made it past her mother and down to the skating rink and skated the day away as planned. It did not end so well though. When Margaret got home that night, the hives had worsened and she was out of school for a couple of days afterward. Her mother wasn't too pleased when she found out that we had known about the hives before we went skating.

During the summer months in Madaket, we spent endless hours at the beach. Curfew was usually at sunset. We got to meet all kinds of new people at the beach, visitors from all over. We usually spent the nights out in a tent in the yard and would have quite the time being scared by Margaret's cousins after we had settled down for the night.

Those days were so simple and so much fun. There were times when we would think how boring life was on the island. We just didn't realize what a wonderful life we had. It was a combination of island living and the times that we were living in. So much would change over the next three decades.

I was around ten when I started smoking and drinking. I don't really know what possessed me to sneak some cigarettes and a beer from the fridge, but I did. Most of the kids in my class had started experimenting too. It was an age of experimentation. "Hippies" were now regular summer visitors on the island. A new generation was being born, it was the age of sex, drugs, and rock and roll. We were all part of it.

*

Times were changing all over our country in the '60s and it was no different on Nantucket. There were protests about the war, and concerts on the beach.

Nantucket has probably one of the strongest "rumor mills" I have ever experienced. At one point, a rumor surfaced that the Beatles were on the island. All of my friends and I who just adored the Beatles plotted for hours on end how to spot them, staking out likely places where we thought they might be staying in hopes of merely catching a glimpse of these superstars. We never found them, but it was great – if frustrating – fun.

My Aunt Louise and Uncle Jim were also constants in my life during my adolescence. I found myself staying with them on many occasions due to Annie or Arnie's repeated admissions to hospitals or detox centers. Thank God for Aunt Louise. She taught me many things that I probably would have never learned had it not been for her.

Aunt Louise also had two homes, a guest house that she oper-
ated during the summer months and her winter home in Sconset.
During the winter months when I spent time living with them in
Sconset, Uncle Jim and I would make crafts together. I remember
the string sailboat that we made together. It was one of my first
big accomplishments, and I was so proud of it. It sure kept us
busy on weekends.

I loved Aunt Louise and Uncle Jim. The times that I spent with
them seemed so normal and so free from the chaotic home life
that I had come to know. There were regular bedtime hours, rules
and expectations.

I had a hard time with the rules, at first, especially Aunt
Louise's rule about finishing the food on your plate and being
grateful for the food that you were given. She had been in the
army or the navy – I forget which. But, when she offered you food
and you turned up your nose or said "yuck," she would plunk a
big scoop on your plate, which she referred to as an "automatic
no-thank-you-helping." If you didn't eat it, you found it neatly
wrapped and waiting for you to finish it the next day.

I learned this lesson the hard way. She liked to serve this stuff
that she and Uncle Jim would chuckle about. They called it
"S.O.S." Once I asked Uncle Jim, "What is S.O.S., anyway?" With
the usual chuckle, he said "Shit on a shingle. That's what they
called it in the army." So I still didn't know what I was in for until it
was dished onto my plate. I made an instant face when I saw it
and, of course, got the big no-thank-you-helping. I pushed it from
one side of the plate to the other and spread it around a bit, trying to make it look like I had attempted to eat it. Nothing was
said when I announced I was finished. Aunt Louise just smiled

and said, "All right, then." I awoke the next morning to find that S.O.S. waiting for me at breakfast. It took me quite a while to finish my breakfast that day.

I got my first job at the five-and-ten as soon as I was of working age. I stocked shelves, and then graduated to operating the cash register. I had other summer jobs throughout my teenage years, working at the local Clam Shanty and also at Aunt Louise's guest house, baby sitting and making up the rooms.

Aunt Louise was very fussy about how you made the beds. She taught me how to make a perfect "hospital corner." If the corner wasn't just so, she would follow me and unmake every bed for me to do over. I hated it when this happened, but again it was her way of teaching me. Nobody had ever taken the time to teach me much of anything domestic. Had she not stepped up to do it, I probably would never have learned many of these kinds of things. She taught me well; I still make those hospital corners today.

We also spent many days cooking. I am grateful for all of the cooking that I learned while staying with my aunt. We used to make date filled cookies, banana cream pie and lots of other tasty treats.

I was earning a good amount of money over the summer months, which I used to buy clothes for school in the fall. It gave me spending money for other things, too, like buying beer and drugs. When I drank or came home late, Annie was usually oblivious to it and Arnie was usually in bed asleep, so I really didn't have to work too hard to hide things from them. Even during the teenage years, I was not drinking normally. I always made sure I had consumed enough to numb the pain inside. It usually resulted in some fiasco or another. I could really make a fool out of myself.

In high school I became involved in sports, which gave me the motivation to stay involved in school and to try to get good grades. I was an average student, with occasional teachers I so frustrated that they had no choice but to fail me.

Football was the "in thing" during my high school years. I was a cheerleader and this was definitely a good thing for me. Vito, our coach, was like a father to many of us. He said all of those things that you didn't want to hear, but he was also ready with encouragement when needed. I looked up to Vito and definitely gave "my all" when it came to being involved as a cheerleader and supporting our team. We had a good team and we were very competitive. Martha's Vineyard (the other island) and P-Town (as Provincetown was known) were our main foes. Those games were always big and we prepared for the "enemy" with pep rallies and long days of practice.

Vito is still coaching to this day – he is another legend among the locals.

Reflecting back on my childhood on Nantucket, I have no regrets. Life was happening just as it was supposed to be. I was making every mistake that was necessary for me to become my own person.

I applied my heart to what I observed and learned a lesson
from what I saw
Proverbs 24:32

Cycles of life

During my rebellious teenage years, I quickly came to realize that the best way to escape the chaos of my home life was to be there as little as possible. I also became aware that my friends and their families knew all too well what was going on with our family, either commenting to me directly about Annie and Arnie and the way they drank, or whispering about it behind my back, whisperings that eventually made their way back to me.

My boyfriend during my teenage years was Bruce Murray. He and his family did what they could to help me through the tough times. I would even spend time staying at their home when things got rough at home with Annie and Arnie. Just like the days when my mother and father had begun dating, the dysfunction in our home remained the same.

Annie would get really belligerent when she was drinking. She was jealous of how close I had become with the Murray family. She would call them and say rude things about them on the phone. They would dismiss the situation and do what they could to help me get through it.

I really began to hate my life. I fantasized about my father. I daydreamed about what he was like and if I would ever get to meet him. I hadn't heard from him or my family in Ohio at all. I didn't really know why. I only knew what Annie and Arnie has told me, that my dad had only been interested in me when there was the chance of financial gain after my mother's death.

I was just getting into my teenage years, and I was drinking and drugging now, pretty much anything and everything I could get my hands on. I became more and more independent – and more and more self-destructive.

Survivor John Shea and his wife owned a second home on the island and were regular summer visitors on Nantucket. Those who regularly return to the island to spend their summers are soon accepted as an extended part of the community. Although I was completely unaware of their concerns for me as a teenager, John and Felice Shea, it turned out, were keeping an eye on me from the background. They felt connected to me after the accident, and wanted to somehow watch out for me.

One of the most regrettable moments for me as I was preparing to write this book was the day that I tried to contact John Shea. I had particularly hoped to thank him for his efforts the night of the crash. I was deeply saddened to learn that he had passed on in February, 1978. This was when Mrs. Shea shared with me their concerns about my life after the crash. Both she and her husband had thought it was terrible that the two sides of my family had ended up hating each other over the settlement money from the crash. She knew a lot about the situation with the custody battle, as she was close to many of the people in the legal community on Nantucket. Mrs. Shea is a judge in New York.

*

I was like a chameleon. I could fit in with whatever crowd I chose to hang out with. I could fit in with the "jocks" or the "druggies" or even the "red necks." The only group I never felt comfortable with was the crowd that excelled in school. I think their lives always seemed too boring to me. I had definitely become attracted to the people that needed "fixing."

After I got the job at the end of high school, I lived on the island for a few more years. I moved out of the house and got my own place. At least I had some space between myself and Annie and Arnie, though, when there was an episode of drinking, it never took either of them long to find me. Neither seemed able to carry on without the other for support.

Right before my eighteenth birthday, a surprise package showed up in the mail, from my Grandma Young in Ohio. It was the ring, the diamond ring that Joe Lennon had given my mother when she got married. My mom had asked her to hold onto it all those years ago, and she had! I was thrilled when I opened the package and read the letter that was included with it.

My Grandma Young told me about how often she had thought about me and how she had waited for this day. She also told me how she had wanted to stay in touch, but whenever she had written or sent packages when I was younger, they all were returned to her. This made me angry. Why had Annie and Arnie lied to me all of these years?

After I had received the surprise package from Ohio, I ran down to the drugstore where my Uncle Jim worked to tell him all about it. I was shocked to learn that he wasn't surprised at all. He told me that he and my aunt had kept in touch with my father's mother over the years to keep her up-to-date on how I was doing. My aunt and uncle had given Grandma Young my address so she could send me the ring.

Neither Annie, Arnie or I had ever known. Uncle Jim asked me not to let Annie or Arnie know for fear that it would destroy their relationship. I held the secret, knowing he was right.

*

Soon after my eighteenth birthday I was hit with another surprise. The trust fund that had been set up for me as a child was structured so that I could start withdrawing from it when I turned eighteen. For the first time, I received a statement addressed to me. The balance in the bank: $1.18.There it was, plain as day.

This was how Annie and Arnie had been able to be out of work all of these years. They had been making withdrawals from the trust fund whenever they needed the cash. I think my biggest regret was that they had managed to piss this money away with little to show for it, not a house, not a car, nothing. It seemed like such a waste! Then I found myself making excuses for them, justifying it to myself, remembering all that they had given me as I was growing up.

Annie was always trying to make up for the latest bout of drinking and chaos. Whenever she could she handed me money, hoping that this would somehow excuse her behavior, I suppose. She never understood that this wasn't at all what I wanted or needed. I went out and blew the money foolishly. I had never had any care or guidance as to how to be responsible with it.

*

After the ring arrived I contacted my Grandma Young. Before long, she sent me a phone number and address for my father. I can still remember the rush of excitement I felt staring at this information. Finally, one day, I summoned up all of the courage I could find and dialed the number.

"Hello" said a man with a slow southern drawl. "Hi," I said, "this is Cindy, your daughter." "Well, I'll be damned" was the reply from my dad.

We talked for at least an hour, catching up on everything: life, Annie and Arnie, my story, his story, my questions and his answers. I was surprised to learn that he had remarried several years after my mother died. Actually, he remarried a total of five times. With one of the marriages he had two sons, my half brothers. He told me that when that marriage ended things were very hostile and that was the last that he had communicated with either of his sons. I thought about how it might be nice to meet them one day. I have not done so to date, but you never know where your path of life will bring you. That day may still be a possibility. We ended the call with a promise and a plan. I was going to get on a plane to Florida and go see him, as soon as I could possibly arrange it.

For days after that phone call, I told anybody who would listen about the amazing, exciting news, that I had reconnected with my "real" dad. Some would listen and just comment "wow." Others could not begin to comprehend what this all meant to me. It didn't matter. My father, that man who I had thought about and fantasized about for so many years, was now "real," and I was headed to Florida to meet him. Life was going to change for the better. I was sure of it!

*

A week or so after the call I had been able to get together enough money to buy myself a round-trip ticket to Florida. I couldn't wait to start filling in those blank spaces, find the missing jigsaw pieces of my life.

As the plane began making its final approach to land, my mind was reeling with emotions: excitement, anticipation, worry, nervousness. What was this going to be like? Dad would be there waiting for me at the airport in Melbourne. I didn't know what he looked like, and he had no idea what I looked like, but we just believed that we would be able to figure it out when I got there.

As I came off the plane I spotted a man who was standing alone in the waiting area with white boots on, a fisherman type. It was him.

As I got closer, I noticed he also had a black eye. He had gotten into a bit of a brawl recently at one of the bars, he said casually. We hugged and headed to the car. It was a bit odd; neither of us knew what to say at first, but finally we started talking and began to feel more comfortable with each other.

The plan, apparently, was to "show" me around to all of his friends at the bars. He had been telling everybody how he was finally going to meet his daughter. Then he introduced me to the "girlfriend." She seemed nice, but I really wasn't in the mood or the mindset to be getting to know her right now. I wanted time for just me and my dad.

We went to a bar, had a couple of drinks and talked for a while. Then we went to the hotel where we'd be staying. I started to feel more nervous. My dad had gotten a nice room, right on the water in Daytona. He didn't have much money and had probably spent all that he had on it, but it was just one room. I suddenly felt the reality of just how much I didn't know about this man. I certainly wasn't comfortable enough to sleep in the same room with him. Panic started to set in, which I think he sensed, because he quickly suggested we go to the house of one of his friends for a party.

So off we went. I felt a bit relieved, though I was still worried as to how I was going to deal with sleeping accommodations later on. When we arrived at the party there were lots of people there, some were my age and some my dad's. He soon disappeared to a bar with the girlfriend and there I was in the middle of a multitude of people I didn't know. Most of the night I spent fending off the advances of a guy who at first was interesting, but soon proved to be quite annoying.

By the end of the night, I was tired and disillusioned. My fantasy had quickly vanished only to be replaced by a hard, cold dose of reality. As it turned out, my worries about the hotel and the room were for naught, at least. We never returned to the hotel.

My dad finally reappeared in the wee hours of the morning. I told him I wanted to go home on the first available flight. We sat there crying, both of us. We both knew that we had screwed up, but neither of us knew where this was going or what could be done to fix it. My dad tried talking to me, but he really didn't know what to say. All we could do was cry and feel terrible about it.

I left the next day. As the plane lifted off the runway, my thoughts were in total disarray. I replayed the last twenty-four hours. I thought about Annie and Arnie, about my entire life and all that I had come to know and believe and hope. My dad was not the knight in shining armor who could save me from this crazy life I was living. He was just as screwed up as everyone else in my life. I also felt a connection to Arnie that I had failed to recognize until now. *He* was "my Dad" – the man who, for better or worse, had raised me and loved me. He had been there for me.

When I returned from Florida, I quickly looked for Arnie, eager to let him know the revelation that I had from the whole experience. I wanted to tell him just how much I loved him and appreciated all that he had done for me. I found Arnie at the Anglers' Club, we sat on the porch that day and talked about a lot. Arnie was relieved that I had made it home safely – happy that I came home.

I also realized that my life would have been just as screwed up or even worse had my "real" father raised me. I had just had one of life's toughest lessons, and I think it was what spun me off into several years of self-destruction. I was angry, and the only person I had to take it out on was me.

I drank and drugged heavily over the next several years. I met a guy, married him and ended up in a totally dysfunctional relationship. We were heavily involved with drugs and partying, never bothering to take care of anything or each other. The marriage, which lasted about five years, ended in divorce.

I stayed in contact with everyone, Annie, Arnie, my dad, my Grandmother Young. Grandma Young was financially well off. She gave me a large sum of cash for my wedding, which we promptly blew. I am ashamed to say that I felt like "she owed me." How callous of me, but I was very sick, both spiritually and emotionally, at this point in time.

When I got divorced, she bailed me out again, giving me money to get a lawyer and get myself out of the mess I had gotten myself in to.

*

Soon after the divorce, I met another guy, the man that I am married to today. I didn't like him in the beginning, sensing that he too was an alcoholic. But there was that attraction, like a magnet. If there was a sick, alcoholic single man, I was certainly, if unconsciously, attracted to him. The sicker they were, the more exciting they seemed to be. I guess I was a "very sick picker" – I was psychologically incapable of choosing a healthy relationship. Empty inside, I was deep in the throes of my own personal struggle.

I think it was also a rebound type of situation; I was really feeling like a looser over my failed first marriage. I didn't feel worthy of anyone's love. I linked up with the guy and we continued in our world of insanity, drinking, drugging, and partying.

It was total and complete chaotic behavior. I had slipped into the life pattern that I had grown up with as a child and I didn't even see it happen. I guess I thought I was smarter than that, that I would never let myself fall prey to the disease that I hated most. At the time, I hadn't even grasped that alcoholism is a disease.

Alcoholism was all that I had ever known. All that I had ever wanted and dreamed about was to find a "normal life," whatever that was, but by this time in my life, I wouldn't have known normal if it hit me between the eyes. I was blinded by addiction. I had become Annie – I was an alcoholic. I wasn't ready face this and throw in the towel, however. There had to be more pain, more destruction, and more disasters before I could see and accept the obvious.

My second marriage, just like the first, was a life of ups and downs. We had things, we lost things. We had money, we had no money. We had good times, we had bad times.

Our relationship was beginning to take a turn for the worse when a friend of ours encouraged us to go to an AA meeting. We were just about headed for divorce when this happened. I had moved out and was ready to begin "putting my life together again," for the second time. I knew that something needed to change. I was sick and tired of the chaos and the unpredictable fiascos that I always seemed to end up in. I was in my thirties and getting very weary of the life I was living.

In A.A. there is a saying that the definition of insanity is doing the same thing over and over, expecting different results. I could identify with this, as I had been repeating insane behaviors in my life. I kept thinking that somehow – someway – I could get it right and I would not end up in the same predicament.

Arnie was very sick; he had been diagnosed with lung cancer. He died just before I got sober. The one thing that I am grateful for is that I had stopped drinking right before the last time that I saw him. Annie had called me and told me that if I wanted to see him I needed to get to down to the island quickly. The doctors had not given Arnie much more time. Both he and Annie were living on the Vineyard at the time of Arnie's death. They had been unable to find affordable housing on Nantucket.

I rushed to the island, and was able to spend time with Arnie. The man that I had known, the jolly heavy set guy was now laying there with a frail body that was giving in. We didn't really talk about the grim sentence that the doctors had issued. He knew what the situation was, I knew it, but neither of us wanted to talk about it. I made small talk, telling him that I was trying to stop drinking, thinking that this might raise his spirits.

He told me "don't worry about anything, he was going clamming soon." That was Arnie's way of telling me he knew that the end was near. I cried, but he continued to tell me not to worry, that it would be okay. I left the hospital that day and headed back to the ferry to catch the boat home. I was filled with sadness knowing that this was the last time that I would probably see Arnie. He died two days later.

*

On March 1, 1987, my husband and I went to an AA meeting together at Seminole Point in Sunapee, New Hampshire. I haven't had a drink or a drug since that day. Little did I know the turn my life was about to take; it had never occurred to me that "normal" could be right around the corner. I got active in the program and active in the twelve steps of recovery.

As they say in AA, you have to change everything about your life. Playmates and playgrounds – they all needed to change. I also had to take a very hard look at my childhood. You see, emotionally I had never grown up. I was still that little child, wounded and hurt by all the trauma of my past.

I write this book having celebrated twenty years of sobriety. I still learn new things about myself every day. When I stopped drinking and drugging, I was thirty-one years old. Emotionally, I was probably about twelve. I had stuffed and buried all of my emotion for years upon years.

There is a saying in the AA that with our addictions, it's like a long walk into the woods. For me, it was a walk that took thirty-one years. I like to relate it to being surrounded by the fog that was so thick that night of the crash. I stayed trapped in a similar fog for a large portion of my life.

Through AA, it only took twelve small steps to free me from my prison. Many worry about the things they do in life leading them to hell. Today, I know that since the day I made the choice to clean up my life and stop the insanity, I no longer fear hell. I have already been there.

Tell it to your children, and let your children tell it to their children, and their children to the next generation.
- Joel 1:3

Lessons from life

There are many lessons to learn along the path of life. The lessons I have learned are a direct result of my life's journey.

I really thought that God had left me in those early childhood years. God never left me. He was there by my side, right from the beginning. He had a plan and a purpose that I could not begin to see; I was blinded by my sorrow.

I can relate to that poem about footprints in the sand. God was there from that moment of impact, watching over me. He has carried me though all of the turbulence in my life where I could not see his presence. Today, I know that he is right there, walking right beside me.

In one instant, God granted me a series of gifts that would never have happened had it not been for the accident. For one thing, I was granted a childhood on a beautiful island, growing up in a small community—the lifelong desire of so many.

When I think about fate and whether things are a part of God's great plan, I think that we often don't know the good that comes from sorrow. In 1960, Nantucket Memorial Airport installed a new landing system, a direct result of the crash. Further, the airport expanded in July of that year, adding an air traffic control tower that would allow this small airport to service an ever-increasing flow of air traffic to and from the island.

Another direct outcome of the crash resulted in precedent-setting aviation tort law. In 1958, "*lex loci delitci,*" or "law of the place where the wrong was committed," was the norm in settling cases where the law differed between states. In 1961, however, the New York courts heard the case of Kilberg v. Northeast Airlines. The round trip ticket purchased by Edward Kilberg, one of the deceased on Flight 258, was purchased in New York. On a breach of contract of safe carriage claim, the New York State courts held that since it was a New York contract, they would not enforce Massachusetts' statutory wrongful death limit of $15,000; they would apply the relevant New York State law. This resulted in one of the largest accidental death settlements up to that time. This case study is cited to demonstrate conflicts of law to this day.

Lita Levine recovered and went on to pursue her love of painting. Sad clowns turned into beautiful landscapes as she crossed over from depression to hope.

*

None of these outcomes could have been imagined at the time of the accident.

Time has a way of demonstrating that there is some good to all that is painful. The life I ended up with taught me many lessons. I am sure that would have been the case, no matter which direction my life had taken after the crash. My life could have gone in several directions. But, I have great faith that where I have been and where I am going is exactly as God has planned it.

I was blessed with a life where not everything was handed to me, where nothing came easily. As a result of this I became motivated to achieve. I also became independent and learned to rely on my own abilities early on. Although this got confusing with my addiction, they are still incredibly wonderful assets.

Along the path of life, we grow from all of our relationships. As a child, I could only look at my relationship with Annie and Arnie and see the pain and the faults. Today, I see the many gifts. They did the best that they could with all of the ability that they had. They loved me and nurtured me, and managed to do that through all their own pain and suffering, including the loss of their only child, something no parent should have to endure.

While indulging my own self will and independence, I neglected to see myself sliding into alcoholism. Perhaps this was part of God's plan, or perhaps it was just the result of my own reckless plan. However, the cross that I was given to carry with this disease resulted in the blessings that I cherish most in my life. Had this not happened, I would have not found the gift of sobriety and the wonderful life that I have today.

Annie and Arnie are long gone, as are many referred to in this story. Annie died five years after I got sober. She never was able to enjoy the gift of sobriety and also the spiritual cleansing that is a result of working the twelve steps. Sadly, her death was filled with fear and torment as she entered those final days, wondering what would be in store on the last day.

I was blessed with a child in 1989, a gift that I know God waited to give me until I was ready. One of the things that makes me most proud today is that I have been there for my child, 100%. There have been no days of being too sick to get myself out of bed. I have been present spiritually, mentally and physically. She is one of life's most wonderful blessings and for that I will be eternally grateful.

Alcoholism and addiction taught me that I was spiritually bankrupt. Until I found and came to rely on a power greater than myself, many of life's gifts were out of reach. Once I turned my will and my life over to the care of God, all things became possible.

As far as power over my own life or the lives of others, the only thing that I have power over is my attitude. I can choose to look at what I have and be grateful or I can choose to want more. I can relish in the joy of life or I can wallow in my own self-pity. I can have hope and look toward the future with confidence or I can be consumed by fear of the unknown. Whatever attitude I choose to have, the most important thing is that I am making that choice. For so many years I made the choice to numb myself so that I would have no other choices; I let my addiction make my choices for me. Those days are long behind me. I love making choices! Who would have guessed that this was where that path through the woods was leading. Today I can say that I truly love the life that I have.

I have never established a relationship with my father. He is one of those rolling stones that gather no moss. He rolls into my life and then back out of it for long periods of time. I don't know why, but I can accept that relationship for all that it is and is not. Again, I need to accept that he is doing the best that he can, given where he is in *his* life.

I still have the longing of never having known my mother, but today I have the presence of mind and faith to know that she knows me. I hope that as she is looking down, smiling at the woman that I have become.

*

This ends the story of coming out of the fog. I hope it demonstrates that many things can be overcome through faith and love. I hope that it will encourage you to give your loved ones a special hug and simply say to them, "I love you."

Today, this moment, is all that each and every one of us have. You never know what the next second, minute, hour or day will bring. We all need to appreciate each moment and make the best all that we have – this moment - right now.

I was too young to know about the crash and I don't remember it, but I have lived with it all of my life. In my mind I can see it as if it were yesterday, the pilot, the people and the wreckage.

I still have the longing of never knowing my mother, but I believe she knows me and that she's been there for me all along. When I close my eyes I go back to that tree where they found me so many years ago, I can see her there with her hands stretched out, leading me out of the fog, and I can see myself with my teddy bear in one hand reaching out to her with the other.

Now that I am a mother myself, I know the depth of the bond that exists between a mother and her child. Those words "save the baby" are the words that tell me just how much she loved me. She loved me so much that she put my needs before her own, that is what mothers do!

Rest in peace,

"I love you Mom."

THE END

AFTERWORD

My deepest regrets stem from the things I have never been able to do versus things that I have done. This story is about an emotional trauma. It is about a single event that changed many lives, an event that occurred in an instant. I often catch myself wondering what life would have been like had this day never occurred.

I am curious if there is a deeper sorrow from remembering a loved one who has passed away, or having no memory at all of a person close to you. All of the victims involved in this story were left with memories of their friends or loved ones. I was left with a story and a hole in my heart. Both took years to come to terms with; both will always be there.

I never knew my mother. For me, she is a person in the story, the pretty woman in the pictures. What I know about her is what I have been told. She is a mystery to me as a person. She is someone I can't wait to meet, when I go "home." I hope that all that I have come to believe about the passage from this life to the next, in heaven, includes seeing our loved ones again. For me, that reunion will be the meeting that I have longed for all of my life.

The beginning of this story is just that to me, a story. While I cannot be sure of all of the specific details, I can recall the pieces that I have been told over and over throughout the years. I have no memories of the crash or of the horror that many cannot forget. Many find the story intriguing, myself included.

This story has left me with a deep curiosity about fate and its impact on our lives. Was this meant to be, was it fate? Is there a certain path that I walk that is a part of God's plan?

This story begins in the dark, in the woods, in the fog. There were many years in the early part of my life where faith and a spiritual connection were absent. Those were the years when I was consumed with darkness. I was deep in the woods, deep in the fog. Even though God was with me, looking after me, something that I am clearly aware of today, I could not recognize it then. I was too consumed with my own pain and the worldly things around me.

Through God's grace I found my way out. I believe that this is the real story, the story that God wants me to tell. After the loss of a loved one, we often reflect and wish that we could share just one more embrace. We sometimes long for one more chance to tell those special people in our lives just how much we love them, how much they mean to us.

Just think what the world would be like if we lived each moment with the awareness that life is short and things can change in an instant. Perhaps, we would spend an extra minute to enjoy a beautiful sunset or linger a bit longer in a serene, quiet moment in the woods. Would we take an extra moment to stop what we were doing and assist a loved one with a request instead of replying, "not right now – I'm busy"? Would we part with a meaningful embrace and a lasting kiss instead of a quick hug and peck on the cheek, sometimes just a nod or waive goodbye?

I find myself pondering these thoughts from time to time, but then I get caught up in the hustle and bustle of my fast-paced life, and forget. It's the human experience, I guess.

The notion of writing this book has tempted me over the years. Eventually, it became an overwhelming desire.

I have come to realize that this piece of history is a moment in time etched into the memories of many of my friends, the "locals." Many interesting books have been published about the history and people of Nantucket. Quite a few mention the plane crash and detail varying accounts of what happened on that hot summer night. A disaster of this magnitude has not occurred since. Hopefully, it never will.

It has been 50 years since these events took place, and time has changed the face of the community and the landscape. I no longer reside on the lovely isle of Nantucket, but it will always be a part of me and a part of my heart. I still make regular trips back there. For me, there has never been, and probably never will be, the same sense of "community" that I experienced living on this small island. It was a unique experience, one I will always cherish.

BIBLIOGRAPHY

23 killed, 11 hurt on Nantucket as Northeast Airliner Cracks Up. (1958, August 16). *The Standard Times*.

(1959). *AIRCRAFT ACCIDENT REPORT.* Washington, DC: Civil Aeronautics Board.

Connie and Joe Indio, C.-E. (1958, August 22). Former Automic Commission Head G.E. Dean Is Victim. *The Nantucket Town Crier - Northeast Airlines Plane Crash Special Edition* , p. 12.

CRIER, N. T. (1958, August 22). NORTHEAST AIRLINES PLANE CRASH SPECIAL EDITION. *The Nantucket Town Crier VOL 12 NO. 7* , p. 12.

Day, S. (1958, August 16). Eye Witness describes tragic hospital scene. *The Standard Times*.

EYE WITNESS ACCOUNTS. (1958, August 22). *The Nantucket Town Crier*, p. 12.

Grandparents get custody of Cindy Lou only 18-Months, she lived. (1958, August 22). *The Nantucket Town Crier*, p. 12.

Hagley Museum and Library, W. D. (Director). (1964). *Dupont Show of the Week - "The Patient in Room 601"* [Motion Picture].

Mooney, R. F. (2000). Nantucket Only Yesterday. In R. F. Mooney, *Nantucket Only Yesterday* (p. 294). Nantucket, Massachusetts: Wesco Publishing.

Project, T. R. (1958, August 22). MY DAY . Hyde Park, Massachusetts.

Quinn, A. (1958, August 16). New York Plane Pancakes in fog short of runway. *The Standard Times* .

SENATE CONGRESSIONAL RECORD. (1958). (p. 3). Washington, DC: Public Record.

Strauss, L. L. *Lewis L. Strauss.* Washington, DC.

Tragedy for Vacation-bound Air Travelers. (1958). *Life Magazine* , 100.

UNITED STATES SENATE . (1956 - 1960). *Historic Archives* .

University, T. G. (1958, August 22). *The Eleanor Roosevelt Papers Project.* Retrieved December 2007, from The George Washington University: C:\PERSONAL\book\My Day by Eleanor Roosevelt, August 22, 1958.mht

choughton@metrocast.net

In Memory

✺

John Burnham

Donald Breswick

David Carey

Jeannie Cavin

Arlene Dabek

Eric Bannister

Barbara Dasch

Gordon E. Dean

Barbara Fienerman

Al Foster

Barbara Gillen

Paula Haken

Helen Hereford

William Hinckley

Edward Kilberg

Virginia Marsh

John Pearson

Sidney Rubenfeld

Harvey Schwamm

Jewel Shabell

David Trauth

John Wehmann

Ira Wright

Jacqueline Young

✺